| NAME: |
| MOBILE: |
| EMAIL: |
| ADDRESS: |

LOG BOOK INFORMATION

| Start Date: | End Date: |

TITLE	PAGE
CLIENT CONTACT	3
ESTIMATE LOG	4 – 187
UNDATED CALENDAR	188 – 200

All rights reserved. No part of this book may be reproduced in any form by any electronic or mechanical means including photocopying, recording, or information storage and retrieval without permission in writing from the author.

IF FOUND, PLEASE CONTACT   **OR**

PLEASE LEAVE US A REVIEW ON AMAZON. THANKS!

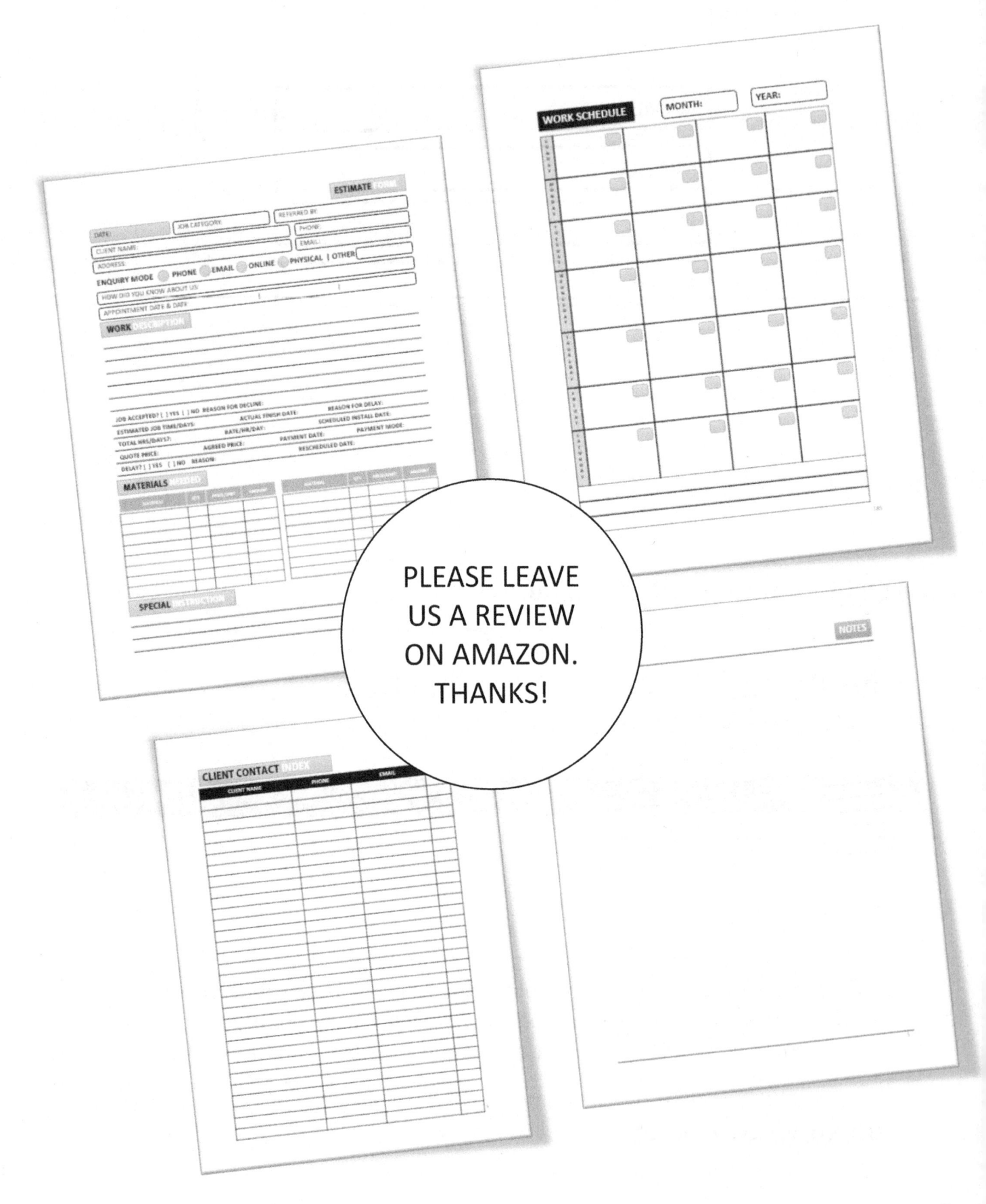

CLIENT CONTACT INDEX

CLIENT NAME	PHONE	EMAIL	PAGE

ESTIMATE FORM

DATE: | JOB CATEGORY: | REFERRED BY:

CLIENT NAME: | PHONE:

ADDRESS: | EMAIL:

ENQUIRY MODE ( ) PHONE ( ) EMAIL ( ) ONLINE ( ) PHYSICAL | OTHER

HOW DID YOU KNOW ABOUT US:

APPOINTMENT DATE & DATE:

WORK DESCRIPTION

_____
_____
_____
_____
_____
_____

JOB ACCEPTED? [ ] YES [ ] NO REASON FOR DECLINE:

ESTIMATED JOB TIME/DAYS: | ACTUAL FINISH DATE: | REASON FOR DELAY:

TOTAL HRS/DAYS?: | RATE/HR/DAY: | SCHEDULED INSTALL DATE:

QUOTE PRICE: | AGREED PRICE: | PAYMENT DATE: | PAYMENT MODE:

DELAY? [ ] YES [ ] NO   REASON: | RESCHEDULED DATE:

MATERIALS NEEDED

MATERIAL	QTY	PRICE/UNIT	AMOUNT	MATERIAL	QTY	PRICE/UNIT	AMOUNT

SPECIAL INSTRUCTION

_____
_____
_____

ESTIMATE NOTES | DIAGRAMS

ESTIMATE FORM

DATE:	JOB CATEGORY:	REFERRED BY:

CLIENT NAME:	PHONE:
ADDRESS:	EMAIL:

ENQUIRY MODE ◯ PHONE ◯ EMAIL ◯ ONLINE ◯ PHYSICAL | OTHER _____

HOW DID YOU KNOW ABOUT US:

APPOINTMENT DATE & DATE: | |

WORK DESCRIPTION

_____
_____
_____
_____
_____
_____

JOB ACCEPTED? [ ] YES [ ] NO REASON FOR DECLINE:

ESTIMATED JOB TIME/DAYS:	ACTUAL FINISH DATE:	REASON FOR DELAY:
TOTAL HRS/DAYS?:	RATE/HR/DAY:	SCHEDULED INSTALL DATE:
QUOTE PRICE:	AGREED PRICE:	PAYMENT DATE:	PAYMENT MODE:

DELAY? [ ] YES [ ] NO REASON: RESCHEDULED DATE:

MATERIALS NEEDED

MATERIAL	QTY	PRICE/UNIT	AMOUNT	MATERIAL	QTY	PRICE/UNIT	AMOUNT

SPECIAL INSTRUCTION

_____
_____
_____

ESTIMATE NOTES | DIAGRAMS

ESTIMATE FORM

DATE:

JOB CATEGORY:

REFERRED BY:

CLIENT NAME:

PHONE:

ADDRESS:

EMAIL:

ENQUIRY MODE ◯ PHONE ◯ EMAIL ◯ ONLINE ◯ PHYSICAL | OTHER

HOW DID YOU KNOW ABOUT US:

APPOINTMENT DATE & DATE: | |

WORK DESCRIPTION

_____
_____
_____
_____
_____
_____
_____

JOB ACCEPTED? [ ] YES [ ] NO REASON FOR DECLINE:

ESTIMATED JOB TIME/DAYS: ACTUAL FINISH DATE: REASON FOR DELAY:

TOTAL HRS/DAYS?:   RATE/HR/DAY: SCHEDULED INSTALL DATE:

QUOTE PRICE: AGREED PRICE: PAYMENT DATE: PAYMENT MODE:

DELAY? [ ] YES [ ] NO REASON: RESCHEDULED DATE:

MATERIALS NEEDED

MATERIAL	QTY	PRICE/UNIT	AMOUNT	MATERIAL	QTY	PRICE/UNIT	AMOUNT

SPECIAL INSTRUCTION

_____
_____
_____

ESTIMATE NOTES | DIAGRAMS

ESTIMATE FORM

| DATE: | JOB CATEGORY: | REFERRED BY: |

| CLIENT NAME: | | PHONE: |
| ADDRESS: | | EMAIL: |

ENQUIRY MODE ○ PHONE ○ EMAIL ○ ONLINE ○ PHYSICAL | OTHER

HOW DID YOU KNOW ABOUT US:

APPOINTMENT DATE & DATE:

WORK DESCRIPTION

_____
_____
_____
_____
_____
_____

JOB ACCEPTED? [ ] YES [ ] NO   REASON FOR DECLINE:

ESTIMATED JOB TIME/DAYS: ACTUAL FINISH DATE: REASON FOR DELAY:

TOTAL HRS/DAYS?:   RATE/HR/DAY: SCHEDULED INSTALL DATE:

QUOTE PRICE: AGREED PRICE: PAYMENT DATE: PAYMENT MODE:

DELAY? [ ] YES   [ ] NO   REASON: RESCHEDULED DATE:

MATERIALS NEEDED

MATERIAL	QTY	PRICE/UNIT	AMOUNT	MATERIAL	QTY	PRICE/UNIT	AMOUNT

SPECIAL INSTRUCTION

_____
_____
_____

ESTIMATE NOTES | DIAGRAMS

ESTIMATE FORM

DATE: JOB CATEGORY: REFERRED BY:

CLIENT NAME: PHONE:

ADDRESS: EMAIL:

ENQUIRY MODE ◯ PHONE ◯ EMAIL ◯ ONLINE ◯ PHYSICAL | OTHER

HOW DID YOU KNOW ABOUT US:

APPOINTMENT DATE & DATE:

WORK DESCRIPTION

_____
_____
_____
_____
_____
_____
_____

JOB ACCEPTED? [ ] YES [ ] NO   REASON FOR DECLINE:

ESTIMATED JOB TIME/DAYS: ACTUAL FINISH DATE: REASON FOR DELAY:

TOTAL HRS/DAYS?:   RATE/HR/DAY: SCHEDULED INSTALL DATE:

QUOTE PRICE: AGREED PRICE: PAYMENT DATE: PAYMENT MODE:

DELAY? [ ] YES [ ] NO   REASON: RESCHEDULED DATE:

MATERIALS NEEDED

MATERIAL	QTY	PRICE/UNIT	AMOUNT	MATERIAL	QTY	PRICE/UNIT	AMOUNT

SPECIAL INSTRUCTION

_____
_____
_____

ESTIMATE NOTES | DIAGRAMS

ESTIMATE FORM

DATE: JOB CATEGORY: REFERRED BY:

CLIENT NAME: PHONE:

ADDRESS: EMAIL:

ENQUIRY MODE   ◯ PHONE   ◯ EMAIL   ◯ ONLINE   ◯ PHYSICAL | OTHER

HOW DID YOU KNOW ABOUT US:

APPOINTMENT DATE & DATE:

WORK DESCRIPTION

_____
_____
_____
_____
_____
_____
_____

JOB ACCEPTED? [ ] YES [ ] NO   REASON FOR DECLINE:

ESTIMATED JOB TIME/DAYS: ACTUAL FINISH DATE: REASON FOR DELAY:

TOTAL HRS/DAYS?:   RATE/HR/DAY: SCHEDULED INSTALL DATE:

QUOTE PRICE: AGREED PRICE: PAYMENT DATE: PAYMENT MODE:

DELAY? [ ] YES   [ ] NO   REASON: RESCHEDULED DATE:

MATERIALS NEEDED

MATERIAL	QTY	PRICE/UNIT	AMOUNT	MATERIAL	QTY	PRICE/UNIT	AMOUNT

SPECIAL INSTRUCTION

_____
_____
_____

ESTIMATE NOTES | DIAGRAMS

ESTIMATE FORM

DATE: _____ JOB CATEGORY: _____ REFERRED BY: _____

CLIENT NAME: _____ PHONE: _____

ADDRESS: _____ EMAIL: _____

ENQUIRY MODE ◯ PHONE ◯ EMAIL ◯ ONLINE ◯ PHYSICAL | OTHER _____

HOW DID YOU KNOW ABOUT US: _____

APPOINTMENT DATE & DATE: _____ | _____ | _____

WORK DESCRIPTION

_____
_____
_____
_____
_____
_____

JOB ACCEPTED? [ ] YES [ ] NO REASON FOR DECLINE: _____

ESTIMATED JOB TIME/DAYS: _____ ACTUAL FINISH DATE: _____ REASON FOR DELAY: _____

TOTAL HRS/DAYS?: _____ RATE/HR/DAY: _____ SCHEDULED INSTALL DATE: _____

QUOTE PRICE: _____ AGREED PRICE: _____ PAYMENT DATE: _____ PAYMENT MODE: _____

DELAY? [ ] YES [ ] NO REASON: _____ RESCHEDULED DATE: _____

MATERIALS NEEDED

MATERIAL	QTY	PRICE/UNIT	AMOUNT	MATERIAL	QTY	PRICE/UNIT	AMOUNT

SPECIAL INSTRUCTION

_____
_____
_____

ESTIMATE NOTES | DIAGRAMS

ESTIMATE FORM

DATE: JOB CATEGORY: REFERRED BY:

CLIENT NAME: PHONE:

ADDRESS: EMAIL:

ENQUIRY MODE ◯ PHONE ◯ EMAIL ◯ ONLINE ◯ PHYSICAL | OTHER

HOW DID YOU KNOW ABOUT US:

APPOINTMENT DATE & DATE:

WORK DESCRIPTION

_____
_____
_____
_____
_____
_____
_____

JOB ACCEPTED? [ ] YES [ ] NO   REASON FOR DECLINE:

ESTIMATED JOB TIME/DAYS: ACTUAL FINISH DATE: REASON FOR DELAY:

TOTAL HRS/DAYS?:   RATE/HR/DAY: SCHEDULED INSTALL DATE:

QUOTE PRICE: AGREED PRICE: PAYMENT DATE: PAYMENT MODE:

DELAY? [ ] YES [ ] NO   REASON: RESCHEDULED DATE:

MATERIALS NEEDED

MATERIAL	QTY	PRICE/UNIT	AMOUNT	MATERIAL	QTY	PRICE/UNIT	AMOUNT

SPECIAL INSTRUCTION

_____
_____
_____

ESTIMATE NOTES | DIAGRAMS

ESTIMATE FORM

DATE: | JOB CATEGORY: | REFERRED BY:

CLIENT NAME: | PHONE:

ADDRESS: | EMAIL:

ENQUIRY MODE ○ PHONE ○ EMAIL ○ ONLINE ○ PHYSICAL | OTHER

HOW DID YOU KNOW ABOUT US:

APPOINTMENT DATE & DATE:

WORK DESCRIPTION

_____
_____
_____
_____
_____
_____
_____

JOB ACCEPTED? [ ] YES [ ] NO REASON FOR DECLINE:

ESTIMATED JOB TIME/DAYS: | ACTUAL FINISH DATE: | REASON FOR DELAY:

TOTAL HRS/DAYS?: | RATE/HR/DAY: | SCHEDULED INSTALL DATE:

QUOTE PRICE: | AGREED PRICE: | PAYMENT DATE: | PAYMENT MODE:

DELAY? [ ] YES [ ] NO REASON: | RESCHEDULED DATE:

MATERIALS NEEDED

MATERIAL	QTY	PRICE/UNIT	AMOUNT	MATERIAL	QTY	PRICE/UNIT	AMOUNT

SPECIAL INSTRUCTION

_____
_____

ESTIMATE NOTES | DIAGRAMS

ESTIMATE FORM

DATE: JOB CATEGORY: REFERRED BY:

CLIENT NAME: PHONE:

ADDRESS: EMAIL:

ENQUIRY MODE ○ PHONE ○ EMAIL ○ ONLINE ○ PHYSICAL | OTHER

HOW DID YOU KNOW ABOUT US:

APPOINTMENT DATE & DATE:

WORK DESCRIPTION

_____
_____
_____
_____
_____
_____

JOB ACCEPTED? [ ] YES [ ] NO REASON FOR DECLINE:

ESTIMATED JOB TIME/DAYS: ACTUAL FINISH DATE: REASON FOR DELAY:

TOTAL HRS/DAYS?:   RATE/HR/DAY: SCHEDULED INSTALL DATE:

QUOTE PRICE: AGREED PRICE: PAYMENT DATE: PAYMENT MODE:

DELAY? [ ] YES [ ] NO   REASON: RESCHEDULED DATE:

MATERIALS NEEDED

MATERIAL	QTY	PRICE/UNIT	AMOUNT	MATERIAL	QTY	PRICE/UNIT	AMOUNT

SPECIAL INSTRUCTION

_____
_____
_____

ESTIMATE NOTES | DIAGRAMS

ESTIMATE FORM

DATE: _____ JOB CATEGORY: _____ REFERRED BY: _____

CLIENT NAME: _____ PHONE: _____

ADDRESS: _____ EMAIL: _____

ENQUIRY MODE ⚪ PHONE ⚪ EMAIL ⚪ ONLINE ⚪ PHYSICAL | OTHER _____

HOW DID YOU KNOW ABOUT US: _____

APPOINTMENT DATE & DATE: _____ | _____ | _____

WORK DESCRIPTION

_____
_____
_____
_____
_____
_____
_____

JOB ACCEPTED? [ ] YES [ ] NO   REASON FOR DECLINE: _____

ESTIMATED JOB TIME/DAYS: _____   ACTUAL FINISH DATE: _____   REASON FOR DELAY: _____

TOTAL HRS/DAYS?: _____   RATE/HR/DAY: _____   SCHEDULED INSTALL DATE: _____

QUOTE PRICE: _____   AGREED PRICE: _____   PAYMENT DATE: _____   PAYMENT MODE: _____

DELAY? [ ] YES [ ] NO   REASON: _____   RESCHEDULED DATE: _____

MATERIALS NEEDED

MATERIAL	QTY	PRICE/UNIT	AMOUNT	MATERIAL	QTY	PRICE/UNIT	AMOUNT

SPECIAL INSTRUCTION

_____
_____
_____

ESTIMATE NOTES | DIAGRAMS

ESTIMATE FORM

DATE: _____ JOB CATEGORY: _____ REFERRED BY: _____

CLIENT NAME: _____ PHONE: _____

ADDRESS: _____ EMAIL: _____

ENQUIRY MODE ◯ PHONE ◯ EMAIL ◯ ONLINE ◯ PHYSICAL | OTHER _____

HOW DID YOU KNOW ABOUT US: _____

APPOINTMENT DATE & DATE: _____

WORK DESCRIPTION

_____
_____
_____
_____
_____
_____

JOB ACCEPTED? [ ] YES [ ] NO   REASON FOR DECLINE:

ESTIMATED JOB TIME/DAYS: _____ ACTUAL FINISH DATE: _____ REASON FOR DELAY: _____

TOTAL HRS/DAYS?: _____ RATE/HR/DAY: _____ SCHEDULED INSTALL DATE: _____

QUOTE PRICE: _____ AGREED PRICE: _____ PAYMENT DATE: _____ PAYMENT MODE: _____

DELAY? [ ] YES [ ] NO   REASON: _____ RESCHEDULED DATE: _____

MATERIALS NEEDED

MATERIAL	QTY	PRICE/UNIT	AMOUNT

MATERIAL	QTY	PRICE/UNIT	AMOUNT

SPECIAL INSTRUCTION

_____
_____
_____

ESTIMATE NOTES | DIAGRAMS

ESTIMATE FORM

DATE: JOB CATEGORY: REFERRED BY:

CLIENT NAME: PHONE:

ADDRESS: EMAIL:

ENQUIRY MODE   ◯ PHONE   ◯ EMAIL   ◯ ONLINE   ◯ PHYSICAL | OTHER

HOW DID YOU KNOW ABOUT US:

APPOINTMENT DATE & DATE:

WORK DESCRIPTION

_____
_____
_____
_____
_____
_____
_____

JOB ACCEPTED? [ ] YES [ ] NO   REASON FOR DECLINE:

ESTIMATED JOB TIME/DAYS: ACTUAL FINISH DATE: REASON FOR DELAY:

TOTAL HRS/DAYS?:   RATE/HR/DAY: SCHEDULED INSTALL DATE:

QUOTE PRICE: AGREED PRICE: PAYMENT DATE: PAYMENT MODE:

DELAY? [ ] YES [ ] NO   REASON: RESCHEDULED DATE:

MATERIALS NEEDED

MATERIAL	QTY	PRICE/UNIT	AMOUNT	MATERIAL	QTY	PRICE/UNIT	AMOUNT

SPECIAL INSTRUCTION

_____
_____
_____

ESTIMATE NOTES | DIAGRAMS

ESTIMATE FORM

DATE: _____ JOB CATEGORY: _____ REFERRED BY: _____

CLIENT NAME: _____ PHONE: _____

ADDRESS: _____ EMAIL: _____

ENQUIRY MODE ○ PHONE ○ EMAIL ○ ONLINE ○ PHYSICAL | OTHER _____

HOW DID YOU KNOW ABOUT US: _____

APPOINTMENT DATE & DATE: _____ | _____ | _____

WORK DESCRIPTION

_____
_____
_____
_____
_____
_____
_____

JOB ACCEPTED? [ ] YES [ ] NO REASON FOR DECLINE:

ESTIMATED JOB TIME/DAYS: ACTUAL FINISH DATE: REASON FOR DELAY:

TOTAL HRS/DAYS?: RATE/HR/DAY: SCHEDULED INSTALL DATE:

QUOTE PRICE: AGREED PRICE: PAYMENT DATE: PAYMENT MODE:

DELAY? [ ] YES [ ] NO REASON: RESCHEDULED DATE:

MATERIALS NEEDED

MATERIAL	QTY	PRICE/UNIT	AMOUNT	MATERIAL	QTY	PRICE/UNIT	AMOUNT

SPECIAL INSTRUCTION

_____
_____
_____

ESTIMATE NOTES | DIAGRAMS

ESTIMATE FORM

DATE: 　　　　　　JOB CATEGORY: 　　　　　　REFERRED BY:

CLIENT NAME: 　　　　　　　　　　　　　PHONE:

ADDRESS: 　　　　　　　　　　　　　　　EMAIL:

ENQUIRY MODE ◯ PHONE ◯ EMAIL ◯ ONLINE ◯ PHYSICAL | OTHER

HOW DID YOU KNOW ABOUT US:

APPOINTMENT DATE & DATE: 　　　　　　|　　　　　　|

WORK DESCRIPTION

_____
_____
_____
_____
_____
_____
_____

JOB ACCEPTED? [ ] YES [ ] NO   REASON FOR DECLINE:

ESTIMATED JOB TIME/DAYS: 　　　　ACTUAL FINISH DATE: 　　　　REASON FOR DELAY:

TOTAL HRS/DAYS?: 　　　　RATE/HR/DAY: 　　　　SCHEDULED INSTALL DATE:

QUOTE PRICE: 　　　AGREED PRICE: 　　　PAYMENT DATE: 　　　PAYMENT MODE:

DELAY? [ ] YES [ ] NO   REASON: 　　　　　　　RESCHEDULED DATE:

MATERIALS NEEDED

MATERIAL	QTY	PRICE/UNIT	AMOUNT	MATERIAL	QTY	PRICE/UNIT	AMOUNT

SPECIAL INSTRUCTION

_____
_____
_____

ESTIMATE NOTES | DIAGRAMS

ESTIMATE FORM

DATE: JOB CATEGORY: REFERRED BY:

CLIENT NAME: PHONE:

ADDRESS: EMAIL:

ENQUIRY MODE ◯ PHONE ◯ EMAIL ◯ ONLINE ◯ PHYSICAL | OTHER

HOW DID YOU KNOW ABOUT US:

APPOINTMENT DATE & DATE:

WORK DESCRIPTION

_____
_____
_____
_____
_____
_____

JOB ACCEPTED? [ ] YES [ ] NO   REASON FOR DECLINE:

ESTIMATED JOB TIME/DAYS: ACTUAL FINISH DATE: REASON FOR DELAY:

TOTAL HRS/DAYS?:   RATE/HR/DAY: SCHEDULED INSTALL DATE:

QUOTE PRICE: AGREED PRICE: PAYMENT DATE: PAYMENT MODE:

DELAY? [ ] YES [ ] NO   REASON: RESCHEDULED DATE:

MATERIALS NEEDED

MATERIAL	QTY	PRICE/UNIT	AMOUNT

MATERIAL	QTY	PRICE/UNIT	AMOUNT

SPECIAL INSTRUCTION

_____
_____
_____

ESTIMATE NOTES | DIAGRAMS

ESTIMATE FORM

DATE: JOB CATEGORY: REFERRED BY:

CLIENT NAME: PHONE:

ADDRESS: EMAIL:

ENQUIRY MODE ◯ PHONE ◯ EMAIL ◯ ONLINE ◯ PHYSICAL | OTHER

HOW DID YOU KNOW ABOUT US:

APPOINTMENT DATE & DATE:

WORK DESCRIPTION

_____
_____
_____
_____
_____
_____
_____

JOB ACCEPTED? [ ] YES [ ] NO   REASON FOR DECLINE:

ESTIMATED JOB TIME/DAYS: ACTUAL FINISH DATE: REASON FOR DELAY:

TOTAL HRS/DAYS?:   RATE/HR/DAY: SCHEDULED INSTALL DATE:

QUOTE PRICE: AGREED PRICE: PAYMENT DATE: PAYMENT MODE:

DELAY? [ ] YES [ ] NO   REASON: RESCHEDULED DATE:

MATERIALS NEEDED

MATERIAL	QTY	PRICE/UNIT	AMOUNT	MATERIAL	QTY	PRICE/UNIT	AMOUNT

SPECIAL INSTRUCTION

_____
_____
_____

ESTIMATE NOTES | DIAGRAMS

ESTIMATE FORM

DATE: 　　　　　JOB CATEGORY: 　　　　　REFERRED BY:

CLIENT NAME: 　　　　　　　　　　　　　PHONE:

ADDRESS: 　　　　　　　　　　　　　　　EMAIL:

ENQUIRY MODE ○ PHONE ○ EMAIL ○ ONLINE ○ PHYSICAL | OTHER _____

HOW DID YOU KNOW ABOUT US:

APPOINTMENT DATE & DATE:

WORK DESCRIPTION

_____
_____
_____
_____
_____
_____
_____

JOB ACCEPTED? [ ] YES [ ] NO   REASON FOR DECLINE:

ESTIMATED JOB TIME/DAYS: 　　　　ACTUAL FINISH DATE: 　　　　REASON FOR DELAY:

TOTAL HRS/DAYS?: 　　　　　　　　RATE/HR/DAY: 　　　　　　　SCHEDULED INSTALL DATE:

QUOTE PRICE: 　　　　AGREED PRICE: 　　　　PAYMENT DATE: 　　　　PAYMENT MODE:

DELAY? [ ] YES [ ] NO REASON: 　　　　　　　　　　RESCHEDULED DATE:

MATERIALS NEEDED

MATERIAL	QTY	PRICE/UNIT	AMOUNT

MATERIAL	QTY	PRICE/UNIT	AMOUNT

SPECIAL INSTRUCTION

_____
_____
_____

ESTIMATE NOTES | DIAGRAMS

ESTIMATE FORM

DATE: _____ JOB CATEGORY: _____ REFERRED BY: _____

CLIENT NAME: _____ PHONE: _____

ADDRESS: _____ EMAIL: _____

ENQUIRY MODE ○ PHONE ○ EMAIL ○ ONLINE ○ PHYSICAL | OTHER _____

HOW DID YOU KNOW ABOUT US: _____

APPOINTMENT DATE & DATE: _____ | _____ | _____

WORK DESCRIPTION

_____
_____
_____
_____
_____
_____
_____

JOB ACCEPTED? [ ] YES [ ] NO REASON FOR DECLINE: _____

ESTIMATED JOB TIME/DAYS: _____ ACTUAL FINISH DATE: _____ REASON FOR DELAY: _____

TOTAL HRS/DAYS?: _____ RATE/HR/DAY: _____ SCHEDULED INSTALL DATE: _____

QUOTE PRICE: _____ AGREED PRICE: _____ PAYMENT DATE: _____ PAYMENT MODE: _____

DELAY? [ ] YES [ ] NO REASON: _____ RESCHEDULED DATE: _____

MATERIALS NEEDED

MATERIAL	QTY	PRICE/UNIT	AMOUNT	MATERIAL	QTY	PRICE/UNIT	AMOUNT

SPECIAL INSTRUCTION

_____
_____
_____

ESTIMATE NOTES | DIAGRAMS

ESTIMATE FORM

DATE: _____ JOB CATEGORY: _____ REFERRED BY: _____

CLIENT NAME: _____ PHONE: _____

ADDRESS: _____ EMAIL: _____

ENQUIRY MODE ○ PHONE ○ EMAIL ○ ONLINE ○ PHYSICAL | OTHER _____

HOW DID YOU KNOW ABOUT US: _____

APPOINTMENT DATE & DATE: _____ | _____ | _____

WORK DESCRIPTION

_____
_____
_____
_____
_____
_____
_____

JOB ACCEPTED? [ ] YES [ ] NO REASON FOR DECLINE: _____

ESTIMATED JOB TIME/DAYS: _____ ACTUAL FINISH DATE: _____ REASON FOR DELAY: _____

TOTAL HRS/DAYS?: _____ RATE/HR/DAY: _____ SCHEDULED INSTALL DATE: _____

QUOTE PRICE: _____ AGREED PRICE: _____ PAYMENT DATE: _____ PAYMENT MODE: _____

DELAY? [ ] YES [ ] NO REASON: _____ RESCHEDULED DATE: _____

MATERIALS NEEDED

MATERIAL	QTY	PRICE/UNIT	AMOUNT	MATERIAL	QTY	PRICE/UNIT	AMOUNT

SPECIAL INSTRUCTION

_____
_____
_____

ESTIMATE NOTES | DIAGRAMS

ESTIMATE FORM

DATE: _____ JOB CATEGORY: _____ REFERRED BY: _____

CLIENT NAME: _____ PHONE: _____

ADDRESS: _____ EMAIL: _____

ENQUIRY MODE ○ PHONE ○ EMAIL ○ ONLINE ○ PHYSICAL | OTHER _____

HOW DID YOU KNOW ABOUT US: _____

APPOINTMENT DATE & DATE: _____ | _____ | _____

WORK DESCRIPTION

_____
_____
_____
_____
_____
_____
_____

JOB ACCEPTED? [ ] YES [ ] NO REASON FOR DECLINE: _____

ESTIMATED JOB TIME/DAYS: _____ ACTUAL FINISH DATE: _____ REASON FOR DELAY: _____

TOTAL HRS/DAYS?: _____ RATE/HR/DAY: _____ SCHEDULED INSTALL DATE: _____

QUOTE PRICE: _____ AGREED PRICE: _____ PAYMENT DATE: _____ PAYMENT MODE: _____

DELAY? [ ] YES [ ] NO REASON: _____ RESCHEDULED DATE: _____

MATERIALS NEEDED

MATERIAL	QTY	PRICE/UNIT	AMOUNT	MATERIAL	QTY	PRICE/UNIT	AMOUNT

SPECIAL INSTRUCTION

_____
_____
_____

ESTIMATE NOTES | DIAGRAMS

ESTIMATE FORM

DATE: JOB CATEGORY: REFERRED BY:

CLIENT NAME: PHONE:

ADDRESS: EMAIL:

ENQUIRY MODE ◯ PHONE ◯ EMAIL ◯ ONLINE ◯ PHYSICAL | OTHER

HOW DID YOU KNOW ABOUT US:

APPOINTMENT DATE & DATE:

WORK DESCRIPTION

_____
_____
_____
_____
_____
_____

JOB ACCEPTED? [ ] YES [ ] NO   REASON FOR DECLINE:

ESTIMATED JOB TIME/DAYS: ACTUAL FINISH DATE: REASON FOR DELAY:

TOTAL HRS/DAYS?:   RATE/HR/DAY: SCHEDULED INSTALL DATE:

QUOTE PRICE: AGREED PRICE: PAYMENT DATE: PAYMENT MODE:

DELAY? [ ] YES [ ] NO   REASON: RESCHEDULED DATE:

MATERIALS NEEDED

MATERIAL	QTY	PRICE/UNIT	AMOUNT	MATERIAL	QTY	PRICE/UNIT	AMOUNT

SPECIAL INSTRUCTION

_____
_____
_____

ESTIMATE NOTES | DIAGRAMS

ESTIMATE FORM

DATE: JOB CATEGORY: REFERRED BY:

CLIENT NAME: PHONE:

ADDRESS: EMAIL:

ENQUIRY MODE   ◯ PHONE   ◯ EMAIL   ◯ ONLINE   ◯ PHYSICAL | OTHER

HOW DID YOU KNOW ABOUT US:

APPOINTMENT DATE & DATE:

WORK DESCRIPTION

_____
_____
_____
_____
_____
_____
_____

JOB ACCEPTED? [ ] YES   [ ] NO   REASON FOR DECLINE:

ESTIMATED JOB TIME/DAYS: ACTUAL FINISH DATE: REASON FOR DELAY:

TOTAL HRS/DAYS?:   RATE/HR/DAY: SCHEDULED INSTALL DATE:

QUOTE PRICE: AGREED PRICE: PAYMENT DATE: PAYMENT MODE:

DELAY? [ ] YES   [ ] NO   REASON: RESCHEDULED DATE:

MATERIALS NEEDED

MATERIAL	QTY	PRICE/UNIT	AMOUNT	MATERIAL	QTY	PRICE/UNIT	AMOUNT

SPECIAL INSTRUCTION

_____
_____
_____

ESTIMATE NOTES | DIAGRAMS

ESTIMATE FORM

DATE: 　　　　　　　　JOB CATEGORY: 　　　　　　　　REFERRED BY:

CLIENT NAME: 　　　　　　　　　　　　　　　　PHONE:

ADDRESS: 　　　　　　　　　　　　　　　　　　EMAIL:

ENQUIRY MODE　◯ PHONE　◯ EMAIL　◯ ONLINE　◯ PHYSICAL　| OTHER

HOW DID YOU KNOW ABOUT US:

APPOINTMENT DATE & DATE:

WORK DESCRIPTION

___
___
___
___
___
___
___

JOB ACCEPTED? [ ] YES [ ] NO REASON FOR DECLINE:

ESTIMATED JOB TIME/DAYS: 　　　　ACTUAL FINISH DATE: 　　　　REASON FOR DELAY:

TOTAL HRS/DAYS?: 　　　　RATE/HR/DAY: 　　　　SCHEDULED INSTALL DATE:

QUOTE PRICE: 　　　　AGREED PRICE: 　　　　PAYMENT DATE: 　　　　PAYMENT MODE:

DELAY? [ ] YES　[ ] NO　REASON: 　　　　　　　　RESCHEDULED DATE:

MATERIALS NEEDED

MATERIAL	QTY	PRICE/UNIT	AMOUNT	MATERIAL	QTY	PRICE/UNIT	AMOUNT

SPECIAL INSTRUCTION

___
___
___

ESTIMATE NOTES | DIAGRAMS

ESTIMATE FORM

DATE: JOB CATEGORY: REFERRED BY:

CLIENT NAME: PHONE:

ADDRESS: EMAIL:

ENQUIRY MODE   ◯ PHONE   ◯ EMAIL   ◯ ONLINE   ◯ PHYSICAL | OTHER

HOW DID YOU KNOW ABOUT US:

APPOINTMENT DATE & DATE:

WORK DESCRIPTION

_____
_____
_____
_____
_____
_____
_____

JOB ACCEPTED? [ ] YES [ ] NO   REASON FOR DECLINE:

ESTIMATED JOB TIME/DAYS: ACTUAL FINISH DATE: REASON FOR DELAY:

TOTAL HRS/DAYS?:   RATE/HR/DAY: SCHEDULED INSTALL DATE:

QUOTE PRICE: AGREED PRICE: PAYMENT DATE: PAYMENT MODE:

DELAY? [ ] YES   [ ] NO   REASON: RESCHEDULED DATE:

MATERIALS NEEDED

MATERIAL	QTY	PRICE/UNIT	AMOUNT	MATERIAL	QTY	PRICE/UNIT	AMOUNT

SPECIAL INSTRUCTION

_____
_____
_____

ESTIMATE NOTES | DIAGRAMS

ESTIMATE FORM

DATE: _____ JOB CATEGORY: _____ REFERRED BY: _____

CLIENT NAME: _____ PHONE: _____

ADDRESS: _____ EMAIL: _____

ENQUIRY MODE ◯ PHONE ◯ EMAIL ◯ ONLINE ◯ PHYSICAL | OTHER _____

HOW DID YOU KNOW ABOUT US: _____

APPOINTMENT DATE & DATE: _____ | _____ | _____

WORK DESCRIPTION

_____
_____
_____
_____
_____
_____

JOB ACCEPTED? [ ] YES [ ] NO REASON FOR DECLINE: _____

ESTIMATED JOB TIME/DAYS: _____ ACTUAL FINISH DATE: _____ REASON FOR DELAY: _____

TOTAL HRS/DAYS?: _____ RATE/HR/DAY: _____ SCHEDULED INSTALL DATE: _____

QUOTE PRICE: _____ AGREED PRICE: _____ PAYMENT DATE: _____ PAYMENT MODE: _____

DELAY? [ ] YES [ ] NO REASON: _____ RESCHEDULED DATE: _____

MATERIALS NEEDED

MATERIAL	QTY	PRICE/UNIT	AMOUNT	MATERIAL	QTY	PRICE/UNIT	AMOUNT

SPECIAL INSTRUCTION

_____
_____
_____

ESTIMATE NOTES | DIAGRAMS

ESTIMATE FORM

DATE: | JOB CATEGORY: | REFERRED BY:

CLIENT NAME: | PHONE:

ADDRESS: | EMAIL:

ENQUIRY MODE ○ PHONE ○ EMAIL ○ ONLINE ○ PHYSICAL | OTHER

HOW DID YOU KNOW ABOUT US:

APPOINTMENT DATE & DATE:

WORK DESCRIPTION

_____
_____
_____
_____
_____
_____

JOB ACCEPTED? [ ] YES [ ] NO   REASON FOR DECLINE:

ESTIMATED JOB TIME/DAYS: | ACTUAL FINISH DATE: | REASON FOR DELAY:

TOTAL HRS/DAYS?: | RATE/HR/DAY: | SCHEDULED INSTALL DATE:

QUOTE PRICE: | AGREED PRICE: | PAYMENT DATE: | PAYMENT MODE:

DELAY? [ ] YES   [ ] NO   REASON: | RESCHEDULED DATE:

MATERIALS NEEDED

MATERIAL	QTY	PRICE/UNIT	AMOUNT	MATERIAL	QTY	PRICE/UNIT	AMOUNT

SPECIAL INSTRUCTION

_____
_____
_____

ESTIMATE NOTES | DIAGRAMS

ESTIMATE FORM

DATE: JOB CATEGORY: REFERRED BY:

CLIENT NAME: PHONE:

ADDRESS: EMAIL:

ENQUIRY MODE ◯ PHONE ◯ EMAIL ◯ ONLINE ◯ PHYSICAL | OTHER

HOW DID YOU KNOW ABOUT US:

APPOINTMENT DATE & DATE:

WORK DESCRIPTION

_____
_____
_____
_____
_____
_____
_____

JOB ACCEPTED? [ ] YES [ ] NO REASON FOR DECLINE:

ESTIMATED JOB TIME/DAYS: ACTUAL FINISH DATE: REASON FOR DELAY:

TOTAL HRS/DAYS?: RATE/HR/DAY: SCHEDULED INSTALL DATE:

QUOTE PRICE: AGREED PRICE: PAYMENT DATE: PAYMENT MODE:

DELAY? [ ] YES [ ] NO REASON: RESCHEDULED DATE:

MATERIALS NEEDED

MATERIAL	QTY	PRICE/UNIT	AMOUNT	MATERIAL	QTY	PRICE/UNIT	AMOUNT

SPECIAL INSTRUCTION

_____
_____
_____

ESTIMATE NOTES | DIAGRAMS

ESTIMATE FORM

DATE: JOB CATEGORY: REFERRED BY:

CLIENT NAME: PHONE:

ADDRESS: EMAIL:

ENQUIRY MODE ◯ PHONE ◯ EMAIL ◯ ONLINE ◯ PHYSICAL | OTHER

HOW DID YOU KNOW ABOUT US:

APPOINTMENT DATE & DATE:

WORK DESCRIPTION

_____
_____
_____
_____
_____
_____
_____

JOB ACCEPTED? [ ] YES [ ] NO   REASON FOR DECLINE:

ESTIMATED JOB TIME/DAYS: ACTUAL FINISH DATE: REASON FOR DELAY:

TOTAL HRS/DAYS?:   RATE/HR/DAY: SCHEDULED INSTALL DATE:

QUOTE PRICE: AGREED PRICE: PAYMENT DATE: PAYMENT MODE:

DELAY? [ ] YES [ ] NO   REASON: RESCHEDULED DATE:

MATERIALS NEEDED

MATERIAL	QTY	PRICE/UNIT	AMOUNT	MATERIAL	QTY	PRICE/UNIT	AMOUNT

SPECIAL INSTRUCTION

_____
_____
_____

ESTIMATE NOTES | DIAGRAMS

ESTIMATE FORM

DATE: | JOB CATEGORY: | REFERRED BY:

CLIENT NAME: | PHONE:

ADDRESS: | EMAIL:

ENQUIRY MODE   ○ PHONE   ○ EMAIL   ○ ONLINE   ○ PHYSICAL | OTHER _____

HOW DID YOU KNOW ABOUT US:

APPOINTMENT DATE & DATE:

WORK DESCRIPTION

_____
_____
_____
_____
_____
_____
_____

JOB ACCEPTED? [ ] YES [ ] NO   REASON FOR DECLINE:

ESTIMATED JOB TIME/DAYS: | ACTUAL FINISH DATE: | REASON FOR DELAY:

TOTAL HRS/DAYS?: | RATE/HR/DAY: | SCHEDULED INSTALL DATE:

QUOTE PRICE: | AGREED PRICE: | PAYMENT DATE: | PAYMENT MODE:

DELAY? [ ] YES [ ] NO   REASON: | RESCHEDULED DATE:

MATERIALS NEEDED

MATERIAL	QTY	PRICE/UNIT	AMOUNT	MATERIAL	QTY	PRICE/UNIT	AMOUNT

SPECIAL INSTRUCTION

_____
_____
_____

ESTIMATE NOTES | DIAGRAMS

ESTIMATE FORM

DATE: | JOB CATEGORY: | REFERRED BY:

CLIENT NAME: | PHONE:

ADDRESS: | EMAIL:

ENQUIRY MODE ◯ PHONE ◯ EMAIL ◯ ONLINE ◯ PHYSICAL | OTHER

HOW DID YOU KNOW ABOUT US:

APPOINTMENT DATE & DATE:

WORK DESCRIPTION

_____
_____
_____
_____
_____
_____

JOB ACCEPTED? [ ] YES [ ] NO   REASON FOR DECLINE:

ESTIMATED JOB TIME/DAYS: **ACTUAL FINISH DATE:**   **REASON FOR DELAY:**

TOTAL HRS/DAYS?:   **RATE/HR/DAY:**   **SCHEDULED INSTALL DATE:**

QUOTE PRICE: **AGREED PRICE:**   **PAYMENT DATE:**   **PAYMENT MODE:**

DELAY? [ ] YES [ ] NO REASON: **RESCHEDULED DATE:**

MATERIALS NEEDED

MATERIAL	QTY	PRICE/UNIT	AMOUNT		MATERIAL	QTY	PRICE/UNIT	AMOUNT

SPECIAL INSTRUCTION

_____
_____
_____

ESTIMATE NOTES | DIAGRAMS

ESTIMATE FORM

DATE: JOB CATEGORY: REFERRED BY:

CLIENT NAME: PHONE:

ADDRESS: EMAIL:

ENQUIRY MODE ◯ PHONE ◯ EMAIL ◯ ONLINE ◯ PHYSICAL | OTHER

HOW DID YOU KNOW ABOUT US:

APPOINTMENT DATE & DATE:

WORK DESCRIPTION

_____
_____
_____
_____
_____
_____
_____

JOB ACCEPTED? [ ] YES [ ] NO   REASON FOR DECLINE:

ESTIMATED JOB TIME/DAYS: ACTUAL FINISH DATE: REASON FOR DELAY:

TOTAL HRS/DAYS?:   RATE/HR/DAY: SCHEDULED INSTALL DATE:

QUOTE PRICE: AGREED PRICE: PAYMENT DATE: PAYMENT MODE:

DELAY? [ ] YES [ ] NO   REASON: RESCHEDULED DATE:

MATERIALS NEEDED

MATERIAL	QTY	PRICE/UNIT	AMOUNT	MATERIAL	QTY	PRICE/UNIT	AMOUNT

SPECIAL INSTRUCTION

_____
_____
_____

ESTIMATE NOTES | DIAGRAMS

ESTIMATE FORM

DATE: 　　　　　　JOB CATEGORY: 　　　　　　REFERRED BY:

CLIENT NAME: 　　　　　　　　　　　　　　　PHONE:

ADDRESS: 　　　　　　　　　　　　　　　　　EMAIL:

ENQUIRY MODE ◯ PHONE ◯ EMAIL ◯ ONLINE ◯ PHYSICAL | OTHER

HOW DID YOU KNOW ABOUT US:

APPOINTMENT DATE & DATE:

WORK DESCRIPTION

_____
_____
_____
_____
_____
_____
_____

JOB ACCEPTED? [ ] YES [ ] NO REASON FOR DECLINE:

ESTIMATED JOB TIME/DAYS: 　　　ACTUAL FINISH DATE: 　　　REASON FOR DELAY:

TOTAL HRS/DAYS?: 　　　RATE/HR/DAY: 　　　SCHEDULED INSTALL DATE:

QUOTE PRICE: 　　　AGREED PRICE: 　　　PAYMENT DATE: 　　　PAYMENT MODE:

DELAY? [ ] YES [ ] NO REASON: 　　　　　　RESCHEDULED DATE:

MATERIALS NEEDED

MATERIAL	QTY	PRICE/UNIT	AMOUNT	MATERIAL	QTY	PRICE/UNIT	AMOUNT

SPECIAL INSTRUCTION

_____
_____
_____

ESTIMATE NOTES | DIAGRAMS

ESTIMATE FORM

DATE: _____ JOB CATEGORY: _____ REFERRED BY: _____

CLIENT NAME: _____ PHONE: _____

ADDRESS: _____ EMAIL: _____

ENQUIRY MODE ◯ PHONE ◯ EMAIL ◯ ONLINE ◯ PHYSICAL | OTHER _____

HOW DID YOU KNOW ABOUT US: _____

APPOINTMENT DATE & DATE: _____ | _____ | _____

WORK DESCRIPTION

_____
_____
_____
_____
_____
_____

JOB ACCEPTED? [ ] YES [ ] NO REASON FOR DECLINE: _____

ESTIMATED JOB TIME/DAYS: _____ ACTUAL FINISH DATE: _____ REASON FOR DELAY: _____

TOTAL HRS/DAYS?: _____ RATE/HR/DAY: _____ SCHEDULED INSTALL DATE: _____

QUOTE PRICE: _____ AGREED PRICE: _____ PAYMENT DATE: _____ PAYMENT MODE: _____

DELAY? [ ] YES [ ] NO REASON: _____ RESCHEDULED DATE: _____

MATERIALS NEEDED

MATERIAL	QTY	PRICE/UNIT	AMOUNT	MATERIAL	QTY	PRICE/UNIT	AMOUNT

SPECIAL INSTRUCTION

_____
_____
_____

ESTIMATE NOTES | DIAGRAMS

ESTIMATE FORM

DATE: JOB CATEGORY: REFERRED BY:

CLIENT NAME: PHONE:

ADDRESS: EMAIL:

ENQUIRY MODE ◯ PHONE ◯ EMAIL ◯ ONLINE ◯ PHYSICAL | OTHER

HOW DID YOU KNOW ABOUT US:

APPOINTMENT DATE & DATE:

WORK DESCRIPTION

_____
_____
_____
_____
_____
_____

JOB ACCEPTED? [ ] YES [ ] NO   REASON FOR DECLINE:

ESTIMATED JOB TIME/DAYS: ACTUAL FINISH DATE: REASON FOR DELAY:

TOTAL HRS/DAYS?:   RATE/HR/DAY: SCHEDULED INSTALL DATE:

QUOTE PRICE: AGREED PRICE: PAYMENT DATE: PAYMENT MODE:

DELAY? [ ] YES [ ] NO   REASON: RESCHEDULED DATE:

MATERIALS NEEDED

MATERIAL	QTY	PRICE/UNIT	AMOUNT	MATERIAL	QTY	PRICE/UNIT	AMOUNT

SPECIAL INSTRUCTION

_____
_____
_____

ESTIMATE NOTES | DIAGRAMS

ESTIMATE FORM

DATE: | JOB CATEGORY: | REFERRED BY:

CLIENT NAME: | PHONE:

ADDRESS: | EMAIL:

ENQUIRY MODE ◯ PHONE ◯ EMAIL ◯ ONLINE ◯ PHYSICAL | OTHER

HOW DID YOU KNOW ABOUT US:

APPOINTMENT DATE & DATE:

WORK DESCRIPTION

_____
_____
_____
_____
_____
_____

JOB ACCEPTED? [ ] YES [ ] NO REASON FOR DECLINE:

ESTIMATED JOB TIME/DAYS: | ACTUAL FINISH DATE: | REASON FOR DELAY:

TOTAL HRS/DAYS?: | RATE/HR/DAY: | SCHEDULED INSTALL DATE:

QUOTE PRICE: | AGREED PRICE: | PAYMENT DATE: | PAYMENT MODE:

DELAY? [ ] YES [ ] NO REASON: | RESCHEDULED DATE:

MATERIALS NEEDED

MATERIAL	QTY	PRICE/UNIT	AMOUNT	MATERIAL	QTY	PRICE/UNIT	AMOUNT

SPECIAL INSTRUCTION

_____
_____
_____

ESTIMATE NOTES | DIAGRAMS

ESTIMATE FORM

DATE:

JOB CATEGORY:

REFERRED BY:

CLIENT NAME:

PHONE:

ADDRESS:

EMAIL:

ENQUIRY MODE ○ PHONE ○ EMAIL ○ ONLINE ○ PHYSICAL | OTHER

HOW DID YOU KNOW ABOUT US:

APPOINTMENT DATE & DATE:

WORK DESCRIPTION

_____
_____
_____
_____
_____
_____
_____

JOB ACCEPTED? [ ] YES [ ] NO REASON FOR DECLINE:

ESTIMATED JOB TIME/DAYS: ACTUAL FINISH DATE: REASON FOR DELAY:

TOTAL HRS/DAYS?:   RATE/HR/DAY: SCHEDULED INSTALL DATE:

QUOTE PRICE: AGREED PRICE: PAYMENT DATE: PAYMENT MODE:

DELAY? [ ] YES [ ] NO REASON: RESCHEDULED DATE:

MATERIALS NEEDED

MATERIAL	QTY	PRICE/UNIT	AMOUNT

MATERIAL	QTY	PRICE/UNIT	AMOUNT

SPECIAL INSTRUCTION

_____
_____
_____

ESTIMATE NOTES | DIAGRAMS

ESTIMATE FORM

DATE: _____ JOB CATEGORY: _____ REFERRED BY: _____

CLIENT NAME: _____ PHONE: _____

ADDRESS: _____ EMAIL: _____

ENQUIRY MODE ○ PHONE ○ EMAIL ○ ONLINE ○ PHYSICAL | OTHER _____

HOW DID YOU KNOW ABOUT US: _____

APPOINTMENT DATE & DATE: _____

WORK DESCRIPTION

_____
_____
_____
_____
_____
_____
_____

JOB ACCEPTED? [ ] YES [ ] NO   REASON FOR DECLINE: _____

ESTIMATED JOB TIME/DAYS: _____   ACTUAL FINISH DATE: _____   REASON FOR DELAY: _____

TOTAL HRS/DAYS?: _____   RATE/HR/DAY: _____   SCHEDULED INSTALL DATE: _____

QUOTE PRICE: _____   AGREED PRICE: _____   PAYMENT DATE: _____   PAYMENT MODE: _____

DELAY? [ ] YES [ ] NO   REASON: _____   RESCHEDULED DATE: _____

MATERIALS NEEDED

MATERIAL	QTY	PRICE/UNIT	AMOUNT	MATERIAL	QTY	PRICE/UNIT	AMOUNT

SPECIAL INSTRUCTION

_____
_____
_____

ESTIMATE NOTES | DIAGRAMS

ESTIMATE FORM

DATE: JOB CATEGORY: REFERRED BY:

CLIENT NAME: PHONE:

ADDRESS: EMAIL:

ENQUIRY MODE   ◯ PHONE   ◯ EMAIL   ◯ ONLINE   ◯ PHYSICAL | OTHER

HOW DID YOU KNOW ABOUT US:

APPOINTMENT DATE & DATE:

WORK DESCRIPTION

_____
_____
_____
_____
_____
_____
_____

JOB ACCEPTED? [ ] YES [ ] NO   REASON FOR DECLINE:

ESTIMATED JOB TIME/DAYS: ACTUAL FINISH DATE: REASON FOR DELAY:

TOTAL HRS/DAYS?:   RATE/HR/DAY: SCHEDULED INSTALL DATE:

QUOTE PRICE: AGREED PRICE: PAYMENT DATE: PAYMENT MODE:

DELAY? [ ] YES [ ] NO   REASON: RESCHEDULED DATE:

MATERIALS NEEDED

MATERIAL	QTY	PRICE/UNIT	AMOUNT	MATERIAL	QTY	PRICE/UNIT	AMOUNT

SPECIAL INSTRUCTION

_____
_____
_____

ESTIMATE NOTES | DIAGRAMS

ESTIMATE FORM

DATE: _____ JOB CATEGORY: _____ REFERRED BY: _____

CLIENT NAME: _____ PHONE: _____

ADDRESS: _____ EMAIL: _____

ENQUIRY MODE ◯ PHONE ◯ EMAIL ◯ ONLINE ◯ PHYSICAL | OTHER _____

HOW DID YOU KNOW ABOUT US: _____

APPOINTMENT DATE & DATE: _____ | _____ | _____

WORK DESCRIPTION

_____
_____
_____
_____
_____
_____
_____

JOB ACCEPTED? [ ] YES [ ] NO   REASON FOR DECLINE: _____

ESTIMATED JOB TIME/DAYS: _____   ACTUAL FINISH DATE: _____   REASON FOR DELAY: _____

TOTAL HRS/DAYS?: _____   RATE/HR/DAY: _____   SCHEDULED INSTALL DATE: _____

QUOTE PRICE: _____   AGREED PRICE: _____   PAYMENT DATE: _____   PAYMENT MODE: _____

DELAY? [ ] YES   [ ] NO   REASON: _____   RESCHEDULED DATE: _____

MATERIALS NEEDED

MATERIAL	QTY	PRICE/UNIT	AMOUNT

MATERIAL	QTY	PRICE/UNIT	AMOUNT

SPECIAL INSTRUCTION

_____
_____
_____

ESTIMATE NOTES | DIAGRAMS

ESTIMATE FORM

| DATE: | JOB CATEGORY: | REFERRED BY: |

| CLIENT NAME: | PHONE: |
| ADDRESS: | EMAIL: |

ENQUIRY MODE ○ PHONE ○ EMAIL ○ ONLINE ○ PHYSICAL | OTHER ____

HOW DID YOU KNOW ABOUT US:

APPOINTMENT DATE & DATE:

WORK DESCRIPTION

_____
_____
_____
_____
_____
_____
_____

JOB ACCEPTED? [ ] YES [ ] NO   REASON FOR DECLINE:

ESTIMATED JOB TIME/DAYS: ACTUAL FINISH DATE: REASON FOR DELAY:

TOTAL HRS/DAYS?:   RATE/HR/DAY: SCHEDULED INSTALL DATE:

QUOTE PRICE: AGREED PRICE: PAYMENT DATE: PAYMENT MODE:

DELAY? [ ] YES   [ ] NO   REASON: RESCHEDULED DATE:

MATERIALS NEEDED

MATERIAL	QTY	PRICE/UNIT	AMOUNT	MATERIAL	QTY	PRICE/UNIT	AMOUNT

SPECIAL INSTRUCTION

_____
_____
_____

ESTIMATE NOTES | DIAGRAMS

ESTIMATE FORM

DATE: 　　　　　JOB CATEGORY: 　　　　　REFERRED BY:

CLIENT NAME: 　　　　　PHONE:

ADDRESS: 　　　　　EMAIL:

ENQUIRY MODE ○ PHONE ○ EMAIL ○ ONLINE ○ PHYSICAL | OTHER

HOW DID YOU KNOW ABOUT US:

APPOINTMENT DATE & DATE:

WORK DESCRIPTION

_____
_____
_____
_____
_____
_____

JOB ACCEPTED? [ ] YES [ ] NO REASON FOR DECLINE:

ESTIMATED JOB TIME/DAYS: 　　　ACTUAL FINISH DATE: 　　　REASON FOR DELAY:

TOTAL HRS/DAYS?: 　　　RATE/HR/DAY: 　　　SCHEDULED INSTALL DATE:

QUOTE PRICE: 　　　AGREED PRICE: 　　　PAYMENT DATE: 　　　PAYMENT MODE:

DELAY? [ ] YES [ ] NO REASON: 　　　RESCHEDULED DATE:

MATERIALS NEEDED

MATERIAL	QTY	PRICE/UNIT	AMOUNT	MATERIAL	QTY	PRICE/UNIT	AMOUNT

SPECIAL INSTRUCTION

_____
_____
_____

ESTIMATE NOTES | DIAGRAMS

ESTIMATE FORM

DATE: | JOB CATEGORY: | REFERRED BY:

CLIENT NAME: | PHONE:

ADDRESS: | EMAIL:

ENQUIRY MODE ○ PHONE ○ EMAIL ○ ONLINE ○ PHYSICAL | OTHER

HOW DID YOU KNOW ABOUT US:

APPOINTMENT DATE & DATE:

WORK DESCRIPTION

_____
_____
_____
_____
_____
_____

JOB ACCEPTED? [ ] YES [ ] NO REASON FOR DECLINE:

ESTIMATED JOB TIME/DAYS: | ACTUAL FINISH DATE: | REASON FOR DELAY:

TOTAL HRS/DAYS?: | RATE/HR/DAY: | SCHEDULED INSTALL DATE:

QUOTE PRICE: | AGREED PRICE: | PAYMENT DATE: | PAYMENT MODE:

DELAY? [ ] YES [ ] NO REASON: | RESCHEDULED DATE:

MATERIALS NEEDED

MATERIAL	QTY	PRICE/UNIT	AMOUNT	MATERIAL	QTY	PRICE/UNIT	AMOUNT

SPECIAL INSTRUCTION

_____
_____
_____

ESTIMATE NOTES | DIAGRAMS

ESTIMATE FORM

DATE: 　　　　　　　JOB CATEGORY: 　　　　　　　REFERRED BY:

CLIENT NAME: 　　　　　　　　　　　　　　　　　　PHONE:

ADDRESS: 　　　　　　　　　　　　　　　　　　　　EMAIL:

ENQUIRY MODE ○ PHONE ○ EMAIL ○ ONLINE ○ PHYSICAL | OTHER

HOW DID YOU KNOW ABOUT US:

APPOINTMENT DATE & DATE:

WORK DESCRIPTION

_____
_____
_____
_____
_____
_____

JOB ACCEPTED? [ ] YES [ ] NO REASON FOR DECLINE:

ESTIMATED JOB TIME/DAYS: 　　　ACTUAL FINISH DATE: 　　　REASON FOR DELAY:

TOTAL HRS/DAYS?: 　　　RATE/HR/DAY: 　　　SCHEDULED INSTALL DATE:

QUOTE PRICE: 　　　AGREED PRICE: 　　　PAYMENT DATE: 　　　PAYMENT MODE:

DELAY? [ ] YES [ ] NO REASON: 　　　RESCHEDULED DATE:

MATERIALS NEEDED

MATERIAL	QTY	PRICE/UNIT	AMOUNT	MATERIAL	QTY	PRICE/UNIT	AMOUNT

SPECIAL INSTRUCTION

_____
_____
_____

ESTIMATE NOTES | DIAGRAMS

ESTIMATE FORM

DATE: _____ JOB CATEGORY: _____ REFERRED BY: _____

CLIENT NAME: _____ PHONE: _____

ADDRESS: _____ EMAIL: _____

ENQUIRY MODE ○ PHONE ○ EMAIL ○ ONLINE ○ PHYSICAL | OTHER _____

HOW DID YOU KNOW ABOUT US: _____

APPOINTMENT DATE & DATE: _____ | _____ | _____

WORK DESCRIPTION

_____
_____
_____
_____
_____
_____

JOB ACCEPTED? [ ] YES [ ] NO REASON FOR DECLINE: _____

ESTIMATED JOB TIME/DAYS: _____ ACTUAL FINISH DATE: _____ REASON FOR DELAY: _____

TOTAL HRS/DAYS?: _____ RATE/HR/DAY: _____ SCHEDULED INSTALL DATE: _____

QUOTE PRICE: _____ AGREED PRICE: _____ PAYMENT DATE: _____ PAYMENT MODE: _____

DELAY? [ ] YES [ ] NO REASON: _____ RESCHEDULED DATE: _____

MATERIALS NEEDED

MATERIAL	QTY	PRICE/UNIT	AMOUNT	MATERIAL	QTY	PRICE/UNIT	AMOUNT

SPECIAL INSTRUCTION

_____
_____
_____

ESTIMATE NOTES | DIAGRAMS

ESTIMATE FORM

DATE: JOB CATEGORY: REFERRED BY:

CLIENT NAME: PHONE:

ADDRESS: EMAIL:

ENQUIRY MODE   ◯ PHONE   ◯ EMAIL   ◯ ONLINE   ◯ PHYSICAL | OTHER

HOW DID YOU KNOW ABOUT US:

APPOINTMENT DATE & DATE:

WORK DESCRIPTION

_____
_____
_____
_____
_____
_____

JOB ACCEPTED? [ ] YES [ ] NO   REASON FOR DECLINE:

ESTIMATED JOB TIME/DAYS: ACTUAL FINISH DATE: REASON FOR DELAY:

TOTAL HRS/DAYS?:   RATE/HR/DAY: SCHEDULED INSTALL DATE:

QUOTE PRICE: AGREED PRICE: PAYMENT DATE: PAYMENT MODE:

DELAY? [ ] YES [ ] NO   REASON: RESCHEDULED DATE:

MATERIALS NEEDED

MATERIAL	QTY	PRICE/UNIT	AMOUNT	MATERIAL	QTY	PRICE/UNIT	AMOUNT

SPECIAL INSTRUCTION

_____
_____
_____

ESTIMATE NOTES | DIAGRAMS

ESTIMATE FORM

DATE: JOB CATEGORY: REFERRED BY:

CLIENT NAME: PHONE:

ADDRESS: EMAIL:

ENQUIRY MODE ◯ PHONE ◯ EMAIL ◯ ONLINE ◯ PHYSICAL | OTHER

HOW DID YOU KNOW ABOUT US:

APPOINTMENT DATE & DATE:

WORK DESCRIPTION

_____
_____
_____
_____
_____
_____

JOB ACCEPTED? [ ] YES [ ] NO   REASON FOR DECLINE:

ESTIMATED JOB TIME/DAYS: ACTUAL FINISH DATE: REASON FOR DELAY:

TOTAL HRS/DAYS?:   RATE/HR/DAY: SCHEDULED INSTALL DATE:

QUOTE PRICE: AGREED PRICE: PAYMENT DATE: PAYMENT MODE:

DELAY? [ ] YES [ ] NO   REASON: RESCHEDULED DATE:

MATERIALS NEEDED

MATERIAL	QTY	PRICE/UNIT	AMOUNT	MATERIAL	QTY	PRICE/UNIT	AMOUNT

SPECIAL INSTRUCTION

_____
_____
_____

ESTIMATE NOTES | DIAGRAMS

ESTIMATE FORM

DATE: ⬜ JOB CATEGORY: ⬜ REFERRED BY: ⬜

CLIENT NAME: ⬜ PHONE: ⬜

ADDRESS: ⬜ EMAIL: ⬜

ENQUIRY MODE ⚪ PHONE ⚪ EMAIL ⚪ ONLINE ⚪ PHYSICAL | OTHER ⬜

HOW DID YOU KNOW ABOUT US:

APPOINTMENT DATE & DATE:

WORK DESCRIPTION

_____
_____
_____
_____
_____
_____

JOB ACCEPTED? [ ] YES [ ] NO REASON FOR DECLINE:

ESTIMATED JOB TIME/DAYS: ACTUAL FINISH DATE: REASON FOR DELAY:

TOTAL HRS/DAYS?:   RATE/HR/DAY: SCHEDULED INSTALL DATE:

QUOTE PRICE: AGREED PRICE: PAYMENT DATE: PAYMENT MODE:

DELAY? [ ] YES   [ ] NO   REASON: RESCHEDULED DATE:

MATERIALS NEEDED

MATERIAL	QTY	PRICE/UNIT	AMOUNT

MATERIAL	QTY	PRICE/UNIT	AMOUNT

SPECIAL INSTRUCTION

_____
_____
_____

ESTIMATE NOTES | DIAGRAMS

ESTIMATE FORM

DATE: JOB CATEGORY: REFERRED BY:

CLIENT NAME: PHONE:

ADDRESS: EMAIL:

ENQUIRY MODE   ○ PHONE   ○ EMAIL   ○ ONLINE   ○ PHYSICAL | OTHER

HOW DID YOU KNOW ABOUT US:

APPOINTMENT DATE & DATE:

WORK DESCRIPTION

_____
_____
_____
_____
_____
_____
_____

JOB ACCEPTED? [ ] YES [ ] NO   REASON FOR DECLINE:

ESTIMATED JOB TIME/DAYS: ACTUAL FINISH DATE: REASON FOR DELAY:

TOTAL HRS/DAYS?:   RATE/HR/DAY: SCHEDULED INSTALL DATE:

QUOTE PRICE: AGREED PRICE: PAYMENT DATE: PAYMENT MODE:

DELAY? [ ] YES [ ] NO   REASON: RESCHEDULED DATE:

MATERIALS NEEDED

MATERIAL	QTY	PRICE/UNIT	AMOUNT	MATERIAL	QTY	PRICE/UNIT	AMOUNT

SPECIAL INSTRUCTION

_____
_____
_____

ESTIMATE NOTES | DIAGRAMS

ESTIMATE FORM

DATE: _____ JOB CATEGORY: _____ REFERRED BY: _____

CLIENT NAME: _____ PHONE: _____

ADDRESS: _____ EMAIL: _____

ENQUIRY MODE ◯ PHONE ◯ EMAIL ◯ ONLINE ◯ PHYSICAL | OTHER _____

HOW DID YOU KNOW ABOUT US: _____

APPOINTMENT DATE & DATE: _____ | _____ | _____

WORK DESCRIPTION

_____
_____
_____
_____
_____
_____

JOB ACCEPTED? [ ] YES [ ] NO REASON FOR DECLINE: _____

ESTIMATED JOB TIME/DAYS: _____ ACTUAL FINISH DATE: _____ REASON FOR DELAY: _____

TOTAL HRS/DAYS?: _____ RATE/HR/DAY: _____ SCHEDULED INSTALL DATE: _____

QUOTE PRICE: _____ AGREED PRICE: _____ PAYMENT DATE: _____ PAYMENT MODE: _____

DELAY? [ ] YES [ ] NO REASON: _____ RESCHEDULED DATE: _____

MATERIALS NEEDED

MATERIAL	QTY	PRICE/UNIT	AMOUNT	MATERIAL	QTY	PRICE/UNIT	AMOUNT

SPECIAL INSTRUCTION

_____
_____
_____

ESTIMATE NOTES | DIAGRAMS

ESTIMATE FORM

DATE: JOB CATEGORY: REFERRED BY:

CLIENT NAME: PHONE:

ADDRESS: EMAIL:

ENQUIRY MODE ○ PHONE ○ EMAIL ○ ONLINE ○ PHYSICAL | OTHER

HOW DID YOU KNOW ABOUT US:

APPOINTMENT DATE & DATE:

WORK DESCRIPTION

___

___

___

___

___

___

JOB ACCEPTED? [ ] YES [ ] NO REASON FOR DECLINE:

ESTIMATED JOB TIME/DAYS: ACTUAL FINISH DATE: REASON FOR DELAY:

TOTAL HRS/DAYS?: RATE/HR/DAY: SCHEDULED INSTALL DATE:

QUOTE PRICE: AGREED PRICE: PAYMENT DATE: PAYMENT MODE:

DELAY? [ ] YES [ ] NO REASON: RESCHEDULED DATE:

MATERIALS NEEDED

MATERIAL	QTY	PRICE/UNIT	AMOUNT	MATERIAL	QTY	PRICE/UNIT	AMOUNT

SPECIAL INSTRUCTION

___

___

___

ESTIMATE NOTES | DIAGRAMS

ESTIMATE FORM

DATE: | JOB CATEGORY: | REFERRED BY:

CLIENT NAME: | PHONE:

ADDRESS: | EMAIL:

ENQUIRY MODE ○ PHONE ○ EMAIL ○ ONLINE ○ PHYSICAL | OTHER

HOW DID YOU KNOW ABOUT US:

APPOINTMENT DATE & DATE:

WORK DESCRIPTION

_____
_____
_____
_____
_____
_____

JOB ACCEPTED? [ ] YES [ ] NO REASON FOR DECLINE:

ESTIMATED JOB TIME/DAYS: | ACTUAL FINISH DATE: | REASON FOR DELAY:

TOTAL HRS/DAYS?: | RATE/HR/DAY: | SCHEDULED INSTALL DATE:

QUOTE PRICE: | AGREED PRICE: | PAYMENT DATE: | PAYMENT MODE:

DELAY? [ ] YES [ ] NO REASON: | RESCHEDULED DATE:

MATERIALS NEEDED

MATERIAL	QTY	PRICE/UNIT	AMOUNT	MATERIAL	QTY	PRICE/UNIT	AMOUNT

SPECIAL INSTRUCTION

_____
_____
_____

ESTIMATE NOTES | DIAGRAMS

ESTIMATE FORM

DATE: _____ JOB CATEGORY: _____ REFERRED BY: _____

CLIENT NAME: _____ PHONE: _____

ADDRESS: _____ EMAIL: _____

ENQUIRY MODE ◯ PHONE ◯ EMAIL ◯ ONLINE ◯ PHYSICAL | OTHER _____

HOW DID YOU KNOW ABOUT US: _____

APPOINTMENT DATE & DATE: _____ | _____ | _____

WORK DESCRIPTION

_____
_____
_____
_____
_____
_____

JOB ACCEPTED? [ ] YES [ ] NO   REASON FOR DECLINE: _____

ESTIMATED JOB TIME/DAYS: _____ ACTUAL FINISH DATE: _____ REASON FOR DELAY: _____

TOTAL HRS/DAYS?: _____ RATE/HR/DAY: _____ SCHEDULED INSTALL DATE: _____

QUOTE PRICE: _____ AGREED PRICE: _____ PAYMENT DATE: _____ PAYMENT MODE: _____

DELAY? [ ] YES [ ] NO   REASON: _____   RESCHEDULED DATE: _____

MATERIALS NEEDED

MATERIAL	QTY	PRICE/UNIT	AMOUNT	MATERIAL	QTY	PRICE/UNIT	AMOUNT

SPECIAL INSTRUCTION

_____
_____
_____

ESTIMATE NOTES | DIAGRAMS

ESTIMATE FORM

DATE: JOB CATEGORY: REFERRED BY:

CLIENT NAME: PHONE:

ADDRESS: EMAIL:

ENQUIRY MODE ○ PHONE ○ EMAIL ○ ONLINE ○ PHYSICAL | OTHER

HOW DID YOU KNOW ABOUT US:

APPOINTMENT DATE & DATE:

WORK DESCRIPTION

_____
_____
_____
_____
_____
_____
_____

JOB ACCEPTED? [ ] YES [ ] NO   REASON FOR DECLINE:

ESTIMATED JOB TIME/DAYS: ACTUAL FINISH DATE: REASON FOR DELAY:

TOTAL HRS/DAYS?:   RATE/HR/DAY: SCHEDULED INSTALL DATE:

QUOTE PRICE: AGREED PRICE: PAYMENT DATE: PAYMENT MODE:

DELAY? [ ] YES [ ] NO   REASON: RESCHEDULED DATE:

MATERIALS NEEDED

MATERIAL	QTY	PRICE/UNIT	AMOUNT	MATERIAL	QTY	PRICE/UNIT	AMOUNT

SPECIAL INSTRUCTION

_____
_____
_____

ESTIMATE NOTES | DIAGRAMS

ESTIMATE FORM

DATE: | JOB CATEGORY: | REFERRED BY:

CLIENT NAME: | PHONE:

ADDRESS: | EMAIL:

ENQUIRY MODE ○ PHONE ○ EMAIL ○ ONLINE ○ PHYSICAL | OTHER

HOW DID YOU KNOW ABOUT US:

APPOINTMENT DATE & DATE:

WORK DESCRIPTION

_____
_____
_____
_____
_____
_____

JOB ACCEPTED? [ ] YES [ ] NO   REASON FOR DECLINE:

ESTIMATED JOB TIME/DAYS: | ACTUAL FINISH DATE: | REASON FOR DELAY:

TOTAL HRS/DAYS?: | RATE/HR/DAY: | SCHEDULED INSTALL DATE:

QUOTE PRICE: | AGREED PRICE: | PAYMENT DATE: | PAYMENT MODE:

DELAY? [ ] YES [ ] NO   REASON: | RESCHEDULED DATE:

MATERIALS NEEDED

MATERIAL	QTY	PRICE/UNIT	AMOUNT	MATERIAL	QTY	PRICE/UNIT	AMOUNT

SPECIAL INSTRUCTION

_____
_____
_____

ESTIMATE NOTES | DIAGRAMS

ESTIMATE FORM

DATE: JOB CATEGORY: REFERRED BY:

CLIENT NAME: PHONE:

ADDRESS: EMAIL:

ENQUIRY MODE ◯ PHONE ◯ EMAIL ◯ ONLINE ◯ PHYSICAL | OTHER

HOW DID YOU KNOW ABOUT US:

APPOINTMENT DATE & DATE:

WORK DESCRIPTION

_____
_____
_____
_____
_____
_____

JOB ACCEPTED? [ ] YES [ ] NO REASON FOR DECLINE:

ESTIMATED JOB TIME/DAYS: ACTUAL FINISH DATE: REASON FOR DELAY:

TOTAL HRS/DAYS?: RATE/HR/DAY: SCHEDULED INSTALL DATE:

QUOTE PRICE: AGREED PRICE: PAYMENT DATE: PAYMENT MODE:

DELAY? [ ] YES [ ] NO REASON: RESCHEDULED DATE:

MATERIALS NEEDED

MATERIAL	QTY	PRICE/UNIT	AMOUNT	MATERIAL	QTY	PRICE/UNIT	AMOUNT

SPECIAL INSTRUCTION

_____
_____
_____

ESTIMATE NOTES | DIAGRAMS

ESTIMATE FORM

DATE: JOB CATEGORY: REFERRED BY:

CLIENT NAME: PHONE:

ADDRESS: EMAIL:

ENQUIRY MODE ○ PHONE ○ EMAIL ○ ONLINE ○ PHYSICAL | OTHER

HOW DID YOU KNOW ABOUT US:

APPOINTMENT DATE & DATE:

WORK DESCRIPTION

_____
_____
_____
_____
_____
_____

JOB ACCEPTED? [ ] YES [ ] NO   REASON FOR DECLINE:

ESTIMATED JOB TIME/DAYS: ACTUAL FINISH DATE: REASON FOR DELAY:

TOTAL HRS/DAYS?:   RATE/HR/DAY: SCHEDULED INSTALL DATE:

QUOTE PRICE: AGREED PRICE: PAYMENT DATE: PAYMENT MODE:

DELAY? [ ] YES [ ] NO   REASON: RESCHEDULED DATE:

MATERIALS NEEDED

MATERIAL	QTY	PRICE/UNIT	AMOUNT	MATERIAL	QTY	PRICE/UNIT	AMOUNT

SPECIAL INSTRUCTION

_____
_____
_____

ESTIMATE NOTES | DIAGRAMS

ESTIMATE FORM

DATE: JOB CATEGORY: REFERRED BY:

CLIENT NAME: PHONE:

ADDRESS: EMAIL:

ENQUIRY MODE ◯ PHONE ◯ EMAIL ◯ ONLINE ◯ PHYSICAL | OTHER

HOW DID YOU KNOW ABOUT US:

APPOINTMENT DATE & DATE:

WORK DESCRIPTION

_____
_____
_____
_____
_____
_____

JOB ACCEPTED? [ ] YES [ ] NO   REASON FOR DECLINE:

ESTIMATED JOB TIME/DAYS: ACTUAL FINISH DATE: REASON FOR DELAY:

TOTAL HRS/DAYS?:   RATE/HR/DAY: SCHEDULED INSTALL DATE:

QUOTE PRICE: AGREED PRICE: PAYMENT DATE: PAYMENT MODE:

DELAY? [ ] YES [ ] NO   REASON: RESCHEDULED DATE:

MATERIALS NEEDED

MATERIAL	QTY	PRICE/UNIT	AMOUNT	MATERIAL	QTY	PRICE/UNIT	AMOUNT

SPECIAL INSTRUCTION

_____
_____
_____

ESTIMATE NOTES | DIAGRAMS

ESTIMATE FORM

DATE: JOB CATEGORY: REFERRED BY:

CLIENT NAME: PHONE:

ADDRESS: EMAIL:

ENQUIRY MODE ◯ PHONE ◯ EMAIL ◯ ONLINE ◯ PHYSICAL | OTHER

HOW DID YOU KNOW ABOUT US:

APPOINTMENT DATE & DATE:

WORK DESCRIPTION

_____
_____
_____
_____
_____
_____

JOB ACCEPTED? [ ] YES [ ] NO   REASON FOR DECLINE:

ESTIMATED JOB TIME/DAYS: ACTUAL FINISH DATE: REASON FOR DELAY:

TOTAL HRS/DAYS?:   RATE/HR/DAY: SCHEDULED INSTALL DATE:

QUOTE PRICE: AGREED PRICE: PAYMENT DATE: PAYMENT MODE:

DELAY? [ ] YES [ ] NO   REASON: RESCHEDULED DATE:

MATERIALS NEEDED

MATERIAL	QTY	PRICE/UNIT	AMOUNT	MATERIAL	QTY	PRICE/UNIT	AMOUNT

SPECIAL INSTRUCTION

_____
_____
_____

ESTIMATE NOTES | DIAGRAMS

ESTIMATE FORM

DATE: _____ **JOB CATEGORY:** _____ **REFERRED BY:** _____

CLIENT NAME: _____ **PHONE:** _____

ADDRESS: _____ **EMAIL:** _____

ENQUIRY MODE ◯ PHONE ◯ EMAIL ◯ ONLINE ◯ PHYSICAL | OTHER _____

HOW DID YOU KNOW ABOUT US: _____

APPOINTMENT DATE & DATE: _____

WORK DESCRIPTION

_____
_____
_____
_____
_____
_____

JOB ACCEPTED? [ ] YES [ ] NO   **REASON FOR DECLINE:** _____

ESTIMATED JOB TIME/DAYS: _____ **ACTUAL FINISH DATE:** _____ **REASON FOR DELAY:** _____

TOTAL HRS/DAYS?: _____ **RATE/HR/DAY:** _____ **SCHEDULED INSTALL DATE:** _____

QUOTE PRICE: _____ **AGREED PRICE:** _____ **PAYMENT DATE:** _____ **PAYMENT MODE:** _____

DELAY? [ ] YES [ ] NO **REASON:** _____ **RESCHEDULED DATE:** _____

MATERIALS NEEDED

MATERIAL	QTY	PRICE/UNIT	AMOUNT

MATERIAL	QTY	PRICE/UNIT	AMOUNT

SPECIAL INSTRUCTION

_____
_____
_____

ESTIMATE NOTES | DIAGRAMS

ESTIMATE FORM

DATE: | JOB CATEGORY: | REFERRED BY:

CLIENT NAME: | PHONE:

ADDRESS: | EMAIL:

ENQUIRY MODE ○ PHONE ○ EMAIL ○ ONLINE ○ PHYSICAL | OTHER

HOW DID YOU KNOW ABOUT US:

APPOINTMENT DATE & DATE:

WORK DESCRIPTION

_____
_____
_____
_____
_____
_____

JOB ACCEPTED? [ ] YES [ ] NO   REASON FOR DECLINE:

ESTIMATED JOB TIME/DAYS: | ACTUAL FINISH DATE: | REASON FOR DELAY:

TOTAL HRS/DAYS?: | RATE/HR/DAY: | SCHEDULED INSTALL DATE:

QUOTE PRICE: | AGREED PRICE: | PAYMENT DATE: | PAYMENT MODE:

DELAY? [ ] YES   [ ] NO   REASON: | RESCHEDULED DATE:

MATERIALS NEEDED

MATERIAL	QTY	PRICE/UNIT	AMOUNT	MATERIAL	QTY	PRICE/UNIT	AMOUNT

SPECIAL INSTRUCTION

_____
_____
_____

ESTIMATE NOTES | DIAGRAMS

ESTIMATE FORM

DATE: | JOB CATEGORY: | REFERRED BY:

CLIENT NAME: | PHONE:

ADDRESS: | EMAIL:

ENQUIRY MODE ○ PHONE ○ EMAIL ○ ONLINE ○ PHYSICAL | OTHER

HOW DID YOU KNOW ABOUT US:

APPOINTMENT DATE & DATE:

WORK DESCRIPTION

_____
_____
_____
_____
_____
_____

JOB ACCEPTED? [ ] YES [ ] NO   REASON FOR DECLINE:

ESTIMATED JOB TIME/DAYS: | ACTUAL FINISH DATE: | REASON FOR DELAY:

TOTAL HRS/DAYS?: | RATE/HR/DAY: | SCHEDULED INSTALL DATE:

QUOTE PRICE: | AGREED PRICE: | PAYMENT DATE: | PAYMENT MODE:

DELAY? [ ] YES [ ] NO   REASON: | RESCHEDULED DATE:

MATERIALS NEEDED

MATERIAL	QTY	PRICE/UNIT	AMOUNT	MATERIAL	QTY	PRICE/UNIT	AMOUNT

SPECIAL INSTRUCTION

_____
_____
_____

ESTIMATE NOTES | DIAGRAMS

ESTIMATE FORM

DATE: | JOB CATEGORY: | REFERRED BY:

CLIENT NAME: | PHONE:

ADDRESS: | EMAIL:

ENQUIRY MODE ○ PHONE ○ EMAIL ○ ONLINE ○ PHYSICAL | OTHER

HOW DID YOU KNOW ABOUT US:

APPOINTMENT DATE & DATE:

WORK DESCRIPTION

_____
_____
_____
_____
_____
_____

JOB ACCEPTED? [ ] YES [ ] NO REASON FOR DECLINE:

ESTIMATED JOB TIME/DAYS: | ACTUAL FINISH DATE: | REASON FOR DELAY:

TOTAL HRS/DAYS?: | RATE/HR/DAY: | SCHEDULED INSTALL DATE:

QUOTE PRICE: | AGREED PRICE: | PAYMENT DATE: | PAYMENT MODE:

DELAY? [ ] YES [ ] NO REASON: | RESCHEDULED DATE:

MATERIALS NEEDED

MATERIAL	QTY	PRICE/UNIT	AMOUNT	MATERIAL	QTY	PRICE/UNIT	AMOUNT

SPECIAL INSTRUCTION

_____
_____
_____

ESTIMATE NOTES | DIAGRAMS

ESTIMATE FORM

DATE: JOB CATEGORY: REFERRED BY:

CLIENT NAME: PHONE:

ADDRESS: EMAIL:

ENQUIRY MODE ◯ PHONE ◯ EMAIL ◯ ONLINE ◯ PHYSICAL | OTHER

HOW DID YOU KNOW ABOUT US:

APPOINTMENT DATE & DATE:

WORK DESCRIPTION

_____
_____
_____
_____
_____
_____

JOB ACCEPTED? [ ] YES [ ] NO REASON FOR DECLINE:

ESTIMATED JOB TIME/DAYS: ACTUAL FINISH DATE: REASON FOR DELAY:

TOTAL HRS/DAYS?: RATE/HR/DAY: SCHEDULED INSTALL DATE:

QUOTE PRICE: AGREED PRICE: PAYMENT DATE: PAYMENT MODE:

DELAY? [ ] YES [ ] NO REASON: RESCHEDULED DATE:

MATERIALS NEEDED

MATERIAL	QTY	PRICE/UNIT	AMOUNT	MATERIAL	QTY	PRICE/UNIT	AMOUNT

SPECIAL INSTRUCTION

_____
_____
_____

ESTIMATE NOTES | DIAGRAMS

ESTIMATE FORM

DATE: JOB CATEGORY: REFERRED BY:

CLIENT NAME: PHONE:

ADDRESS: EMAIL:

ENQUIRY MODE   ◯ PHONE   ◯ EMAIL   ◯ ONLINE   ◯ PHYSICAL | OTHER

HOW DID YOU KNOW ABOUT US:

APPOINTMENT DATE & DATE:

WORK DESCRIPTION

_____
_____
_____
_____
_____
_____
_____

JOB ACCEPTED? [ ] YES [ ] NO   REASON FOR DECLINE:

ESTIMATED JOB TIME/DAYS: ACTUAL FINISH DATE: REASON FOR DELAY:

TOTAL HRS/DAYS?:   RATE/HR/DAY: SCHEDULED INSTALL DATE:

QUOTE PRICE: AGREED PRICE: PAYMENT DATE: PAYMENT MODE:

DELAY? [ ] YES   [ ] NO   REASON: RESCHEDULED DATE:

MATERIALS NEEDED

MATERIAL	QTY	PRICE/UNIT	AMOUNT	MATERIAL	QTY	PRICE/UNIT	AMOUNT

SPECIAL INSTRUCTION

_____
_____
_____

ESTIMATE NOTES | DIAGRAMS

ESTIMATE FORM

DATE: | JOB CATEGORY: | REFERRED BY:

CLIENT NAME: | PHONE:

ADDRESS: | EMAIL:

ENQUIRY MODE ◯ PHONE ◯ EMAIL ◯ ONLINE ◯ PHYSICAL | OTHER

HOW DID YOU KNOW ABOUT US:

APPOINTMENT DATE & DATE:

WORK DESCRIPTION

_____
_____
_____
_____
_____
_____
_____

JOB ACCEPTED? [ ] YES [ ] NO   REASON FOR DECLINE:

ESTIMATED JOB TIME/DAYS: | ACTUAL FINISH DATE: | REASON FOR DELAY:

TOTAL HRS/DAYS?: | RATE/HR/DAY: | SCHEDULED INSTALL DATE:

QUOTE PRICE: | AGREED PRICE: | PAYMENT DATE: | PAYMENT MODE:

DELAY? [ ] YES   [ ] NO   REASON: | RESCHEDULED DATE:

MATERIALS NEEDED

MATERIAL	QTY	PRICE/UNIT	AMOUNT	MATERIAL	QTY	PRICE/UNIT	AMOUNT

SPECIAL INSTRUCTION

_____
_____
_____

ESTIMATE NOTES | DIAGRAMS

ESTIMATE FORM

DATE: JOB CATEGORY: REFERRED BY:

CLIENT NAME: PHONE:

ADDRESS: EMAIL:

ENQUIRY MODE ◯ PHONE ◯ EMAIL ◯ ONLINE ◯ PHYSICAL | OTHER

HOW DID YOU KNOW ABOUT US:

APPOINTMENT DATE & DATE:

WORK DESCRIPTION

_____
_____
_____
_____
_____
_____

JOB ACCEPTED? [ ] YES [ ] NO   REASON FOR DECLINE:

ESTIMATED JOB TIME/DAYS: ACTUAL FINISH DATE: REASON FOR DELAY:

TOTAL HRS/DAYS?:   RATE/HR/DAY: SCHEDULED INSTALL DATE:

QUOTE PRICE: AGREED PRICE: PAYMENT DATE: PAYMENT MODE:

DELAY? [ ] YES [ ] NO   REASON: RESCHEDULED DATE:

MATERIALS NEEDED

MATERIAL	QTY	PRICE/UNIT	AMOUNT	MATERIAL	QTY	PRICE/UNIT	AMOUNT

SPECIAL INSTRUCTION

_____
_____
_____

ESTIMATE NOTES | DIAGRAMS

ESTIMATE FORM

DATE:

JOB CATEGORY:

REFERRED BY:

CLIENT NAME:

PHONE:

ADDRESS:

EMAIL:

ENQUIRY MODE ◯ PHONE ◯ EMAIL ◯ ONLINE ◯ PHYSICAL | OTHER

HOW DID YOU KNOW ABOUT US:

APPOINTMENT DATE & DATE:

WORK DESCRIPTION

_____
_____
_____
_____
_____
_____

JOB ACCEPTED? [ ] YES [ ] NO REASON FOR DECLINE:

ESTIMATED JOB TIME/DAYS: ACTUAL FINISH DATE: REASON FOR DELAY:

TOTAL HRS/DAYS?:   RATE/HR/DAY: SCHEDULED INSTALL DATE:

QUOTE PRICE: AGREED PRICE: PAYMENT DATE: PAYMENT MODE:

DELAY? [ ] YES [ ] NO REASON: RESCHEDULED DATE:

MATERIALS NEEDED

MATERIAL	QTY	PRICE/UNIT	AMOUNT	MATERIAL	QTY	PRICE/UNIT	AMOUNT

SPECIAL INSTRUCTION

_____
_____
_____

ESTIMATE NOTES | DIAGRAMS

ESTIMATE FORM

DATE: JOB CATEGORY: REFERRED BY:

CLIENT NAME: PHONE:

ADDRESS: EMAIL:

ENQUIRY MODE ◯ PHONE ◯ EMAIL ◯ ONLINE ◯ PHYSICAL | OTHER

HOW DID YOU KNOW ABOUT US:

APPOINTMENT DATE & DATE: | |

WORK DESCRIPTION

_____
_____
_____
_____
_____
_____
_____

JOB ACCEPTED? [ ] YES [ ] NO   REASON FOR DECLINE:

ESTIMATED JOB TIME/DAYS: ACTUAL FINISH DATE: REASON FOR DELAY:

TOTAL HRS/DAYS?:   RATE/HR/DAY: SCHEDULED INSTALL DATE:

QUOTE PRICE: AGREED PRICE: PAYMENT DATE: PAYMENT MODE:

DELAY? [ ] YES [ ] NO   REASON: RESCHEDULED DATE:

MATERIALS NEEDED

MATERIAL	QTY	PRICE/UNIT	AMOUNT	MATERIAL	QTY	PRICE/UNIT	AMOUNT

SPECIAL INSTRUCTION

_____
_____
_____

ESTIMATE NOTES | DIAGRAMS

ESTIMATE FORM

DATE: JOB CATEGORY: REFERRED BY:

CLIENT NAME: PHONE:

ADDRESS: EMAIL:

ENQUIRY MODE   ◯ PHONE   ◯ EMAIL   ◯ ONLINE   ◯ PHYSICAL | OTHER

HOW DID YOU KNOW ABOUT US:

APPOINTMENT DATE & DATE:

WORK DESCRIPTION

_____
_____
_____
_____
_____
_____
_____

JOB ACCEPTED? [ ] YES [ ] NO   REASON FOR DECLINE:

ESTIMATED JOB TIME/DAYS: ACTUAL FINISH DATE: REASON FOR DELAY:

TOTAL HRS/DAYS?:   RATE/HR/DAY: SCHEDULED INSTALL DATE:

QUOTE PRICE: AGREED PRICE: PAYMENT DATE: PAYMENT MODE:

DELAY? [ ] YES [ ] NO   REASON: RESCHEDULED DATE:

MATERIALS NEEDED

MATERIAL	QTY	PRICE/UNIT	AMOUNT	MATERIAL	QTY	PRICE/UNIT	AMOUNT

SPECIAL INSTRUCTION

_____
_____
_____

ESTIMATE NOTES | DIAGRAMS

ESTIMATE FORM

DATE: JOB CATEGORY: REFERRED BY:

CLIENT NAME: PHONE:

ADDRESS: EMAIL:

ENQUIRY MODE ◯ PHONE ◯ EMAIL ◯ ONLINE ◯ PHYSICAL | OTHER

HOW DID YOU KNOW ABOUT US:

APPOINTMENT DATE & DATE:

WORK DESCRIPTION

_____
_____
_____
_____
_____
_____
_____

JOB ACCEPTED? [ ] YES [ ] NO   REASON FOR DECLINE:

ESTIMATED JOB TIME/DAYS: ACTUAL FINISH DATE: REASON FOR DELAY:

TOTAL HRS/DAYS?:   RATE/HR/DAY: SCHEDULED INSTALL DATE:

QUOTE PRICE: AGREED PRICE: PAYMENT DATE: PAYMENT MODE:

DELAY? [ ] YES [ ] NO   REASON: RESCHEDULED DATE:

MATERIALS NEEDED

MATERIAL	QTY	PRICE/UNIT	AMOUNT	MATERIAL	QTY	PRICE/UNIT	AMOUNT

SPECIAL INSTRUCTION

_____
_____
_____

ESTIMATE NOTES | DIAGRAMS

ESTIMATE FORM

| DATE: | JOB CATEGORY: | REFERRED BY: |

CLIENT NAME: PHONE:

ADDRESS: EMAIL:

ENQUIRY MODE ◯ PHONE ◯ EMAIL ◯ ONLINE ◯ PHYSICAL | OTHER

HOW DID YOU KNOW ABOUT US:

APPOINTMENT DATE & DATE:

WORK DESCRIPTION

_____
_____
_____
_____
_____
_____

JOB ACCEPTED? [ ] YES [ ] NO   REASON FOR DECLINE:

ESTIMATED JOB TIME/DAYS: ACTUAL FINISH DATE: REASON FOR DELAY:

TOTAL HRS/DAYS?:   RATE/HR/DAY: SCHEDULED INSTALL DATE:

QUOTE PRICE: AGREED PRICE: PAYMENT DATE: PAYMENT MODE:

DELAY? [ ] YES [ ] NO   REASON: RESCHEDULED DATE:

MATERIALS NEEDED

MATERIAL	QTY	PRICE/UNIT	AMOUNT	MATERIAL	QTY	PRICE/UNIT	AMOUNT

SPECIAL INSTRUCTION

_____
_____
_____

ESTIMATE NOTES | DIAGRAMS

ESTIMATE FORM

DATE: JOB CATEGORY: REFERRED BY:

CLIENT NAME: PHONE:

ADDRESS: EMAIL:

ENQUIRY MODE ◯ PHONE ◯ EMAIL ◯ ONLINE ◯ PHYSICAL | OTHER

HOW DID YOU KNOW ABOUT US:

APPOINTMENT DATE & DATE:

WORK DESCRIPTION

_____
_____
_____
_____
_____
_____
_____

JOB ACCEPTED? [ ] YES [ ] NO REASON FOR DECLINE:

ESTIMATED JOB TIME/DAYS: ACTUAL FINISH DATE: REASON FOR DELAY:

TOTAL HRS/DAYS?:   RATE/HR/DAY: SCHEDULED INSTALL DATE:

QUOTE PRICE: AGREED PRICE: PAYMENT DATE: PAYMENT MODE:

DELAY? [ ] YES [ ] NO REASON: RESCHEDULED DATE:

MATERIALS NEEDED

MATERIAL	QTY	PRICE/UNIT	AMOUNT	MATERIAL	QTY	PRICE/UNIT	AMOUNT

SPECIAL INSTRUCTION

_____
_____
_____

ESTIMATE NOTES | DIAGRAMS

ESTIMATE FORM

DATE:

JOB CATEGORY:

REFERRED BY:

CLIENT NAME:

PHONE:

ADDRESS:

EMAIL:

ENQUIRY MODE ○ PHONE ○ EMAIL ○ ONLINE ○ PHYSICAL | OTHER

HOW DID YOU KNOW ABOUT US:

APPOINTMENT DATE & DATE:

WORK DESCRIPTION

_____
_____
_____
_____
_____
_____

JOB ACCEPTED? [ ] YES [ ] NO REASON FOR DECLINE:

ESTIMATED JOB TIME/DAYS: ACTUAL FINISH DATE: REASON FOR DELAY:

TOTAL HRS/DAYS?:   RATE/HR/DAY: SCHEDULED INSTALL DATE:

QUOTE PRICE: AGREED PRICE: PAYMENT DATE: PAYMENT MODE:

DELAY? [ ] YES [ ] NO REASON: RESCHEDULED DATE:

MATERIALS NEEDED

MATERIAL	QTY	PRICE/UNIT	AMOUNT	MATERIAL	QTY	PRICE/UNIT	AMOUNT

SPECIAL INSTRUCTION

_____
_____
_____

ESTIMATE NOTES | DIAGRAMS

ESTIMATE FORM

DATE: JOB CATEGORY: REFERRED BY:

CLIENT NAME: PHONE:

ADDRESS: EMAIL:

ENQUIRY MODE ◯ PHONE ◯ EMAIL ◯ ONLINE ◯ PHYSICAL | OTHER

HOW DID YOU KNOW ABOUT US:

APPOINTMENT DATE & DATE:

WORK DESCRIPTION

_____
_____
_____
_____
_____
_____

JOB ACCEPTED? [ ] YES [ ] NO   REASON FOR DECLINE:

ESTIMATED JOB TIME/DAYS: ACTUAL FINISH DATE: REASON FOR DELAY:

TOTAL HRS/DAYS?:   RATE/HR/DAY: SCHEDULED INSTALL DATE:

QUOTE PRICE: AGREED PRICE: PAYMENT DATE: PAYMENT MODE:

DELAY? [ ] YES [ ] NO   REASON: RESCHEDULED DATE:

MATERIALS NEEDED

MATERIAL	QTY	PRICE/UNIT	AMOUNT	MATERIAL	QTY	PRICE/UNIT	AMOUNT

SPECIAL INSTRUCTION

_____
_____
_____

ESTIMATE NOTES | DIAGRAMS

ESTIMATE FORM

DATE: 　　　　　　JOB CATEGORY: 　　　　　　REFERRED BY:

CLIENT NAME: 　　　　　　　　　　　　　　　PHONE:

ADDRESS: 　　　　　　　　　　　　　　　　　EMAIL:

ENQUIRY MODE　◯ PHONE　◯ EMAIL　◯ ONLINE　◯ PHYSICAL　| OTHER

HOW DID YOU KNOW ABOUT US:

APPOINTMENT DATE & DATE:

WORK DESCRIPTION

_____
_____
_____
_____
_____
_____

JOB ACCEPTED? [ ] YES [ ] NO REASON FOR DECLINE:

ESTIMATED JOB TIME/DAYS: 　　　ACTUAL FINISH DATE: 　　　REASON FOR DELAY:

TOTAL HRS/DAYS?: 　　　RATE/HR/DAY: 　　　SCHEDULED INSTALL DATE:

QUOTE PRICE: 　　　AGREED PRICE: 　　　PAYMENT DATE: 　　　PAYMENT MODE:

DELAY? [ ] YES   [ ] NO   REASON: 　　　　　　RESCHEDULED DATE:

MATERIALS NEEDED

MATERIAL	QTY	PRICE/UNIT	AMOUNT	MATERIAL	QTY	PRICE/UNIT	AMOUNT

SPECIAL INSTRUCTION

_____
_____
_____

ESTIMATE NOTES | DIAGRAMS

ESTIMATE FORM

DATE:

JOB CATEGORY:

REFERRED BY:

CLIENT NAME:

PHONE:

ADDRESS:

EMAIL:

ENQUIRY MODE ○ PHONE ○ EMAIL ○ ONLINE ○ PHYSICAL | OTHER

HOW DID YOU KNOW ABOUT US:

APPOINTMENT DATE & DATE:

WORK DESCRIPTION

_____
_____
_____
_____
_____
_____
_____

JOB ACCEPTED? [ ] YES [ ] NO REASON FOR DECLINE:

ESTIMATED JOB TIME/DAYS: ACTUAL FINISH DATE: REASON FOR DELAY:

TOTAL HRS/DAYS?:   RATE/HR/DAY: SCHEDULED INSTALL DATE:

QUOTE PRICE: AGREED PRICE: PAYMENT DATE: PAYMENT MODE:

DELAY? [ ] YES   [ ] NO   REASON: RESCHEDULED DATE:

MATERIALS NEEDED

MATERIAL	QTY	PRICE/UNIT	AMOUNT

MATERIAL	QTY	PRICE/UNIT	AMOUNT

SPECIAL INSTRUCTION

_____
_____
_____

ESTIMATE NOTES | DIAGRAMS

ESTIMATE FORM

DATE: JOB CATEGORY: REFERRED BY:

CLIENT NAME: PHONE:

ADDRESS: EMAIL:

ENQUIRY MODE ( ) PHONE ( ) EMAIL ( ) ONLINE ( ) PHYSICAL | OTHER _____

HOW DID YOU KNOW ABOUT US:

APPOINTMENT DATE & DATE:

WORK DESCRIPTION

_____
_____
_____
_____
_____
_____
_____

JOB ACCEPTED? [ ] YES [ ] NO   REASON FOR DECLINE:

ESTIMATED JOB TIME/DAYS: ACTUAL FINISH DATE: REASON FOR DELAY:

TOTAL HRS/DAYS?:   RATE/HR/DAY: SCHEDULED INSTALL DATE:

QUOTE PRICE: AGREED PRICE: PAYMENT DATE: PAYMENT MODE:

DELAY? [ ] YES [ ] NO   REASON: RESCHEDULED DATE:

MATERIALS NEEDED

MATERIAL	QTY	PRICE/UNIT	AMOUNT	MATERIAL	QTY	PRICE/UNIT	AMOUNT

SPECIAL INSTRUCTION

_____
_____
_____

ESTIMATE NOTES | DIAGRAMS

ESTIMATE FORM

DATE: JOB CATEGORY: REFERRED BY:

CLIENT NAME: PHONE:

ADDRESS: EMAIL:

ENQUIRY MODE ◯ PHONE ◯ EMAIL ◯ ONLINE ◯ PHYSICAL | OTHER

HOW DID YOU KNOW ABOUT US:

APPOINTMENT DATE & DATE:

WORK DESCRIPTION

_____
_____
_____
_____
_____
_____

JOB ACCEPTED? [ ] YES [ ] NO   REASON FOR DECLINE:

ESTIMATED JOB TIME/DAYS: ACTUAL FINISH DATE: REASON FOR DELAY:

TOTAL HRS/DAYS?:   RATE/HR/DAY: SCHEDULED INSTALL DATE:

QUOTE PRICE: AGREED PRICE: PAYMENT DATE: PAYMENT MODE:

DELAY? [ ] YES [ ] NO   REASON: RESCHEDULED DATE:

MATERIALS NEEDED

MATERIAL	QTY	PRICE/UNIT	AMOUNT	MATERIAL	QTY	PRICE/UNIT	AMOUNT

SPECIAL INSTRUCTION

_____
_____
_____

ESTIMATE NOTES | DIAGRAMS

ESTIMATE FORM

DATE: JOB CATEGORY: REFERRED BY:

CLIENT NAME: PHONE:

ADDRESS: EMAIL:

ENQUIRY MODE ○ PHONE ○ EMAIL ○ ONLINE ○ PHYSICAL | OTHER

HOW DID YOU KNOW ABOUT US:

APPOINTMENT DATE & DATE:

WORK DESCRIPTION

_____
_____
_____
_____
_____
_____
_____

JOB ACCEPTED? [ ] YES [ ] NO   REASON FOR DECLINE:

ESTIMATED JOB TIME/DAYS: ACTUAL FINISH DATE: REASON FOR DELAY:

TOTAL HRS/DAYS?:   RATE/HR/DAY: SCHEDULED INSTALL DATE:

QUOTE PRICE: AGREED PRICE: PAYMENT DATE: PAYMENT MODE:

DELAY? [ ] YES [ ] NO   REASON: RESCHEDULED DATE:

MATERIALS NEEDED

MATERIAL	QTY	PRICE/UNIT	AMOUNT	MATERIAL	QTY	PRICE/UNIT	AMOUNT

SPECIAL INSTRUCTION

_____
_____
_____

ESTIMATE NOTES | DIAGRAMS

ESTIMATE FORM

DATE: JOB CATEGORY: REFERRED BY:

CLIENT NAME: PHONE:

ADDRESS: EMAIL:

ENQUIRY MODE ○ PHONE ○ EMAIL ○ ONLINE ○ PHYSICAL | OTHER

HOW DID YOU KNOW ABOUT US:

APPOINTMENT DATE & DATE:

WORK DESCRIPTION

_____
_____
_____
_____
_____
_____

JOB ACCEPTED? [ ] YES [ ] NO   REASON FOR DECLINE:

ESTIMATED JOB TIME/DAYS: ACTUAL FINISH DATE: REASON FOR DELAY:

TOTAL HRS/DAYS?:   RATE/HR/DAY: SCHEDULED INSTALL DATE:

QUOTE PRICE: AGREED PRICE: PAYMENT DATE: PAYMENT MODE:

DELAY? [ ] YES [ ] NO   REASON: RESCHEDULED DATE:

MATERIALS NEEDED

MATERIAL	QTY	PRICE/UNIT	AMOUNT	MATERIAL	QTY	PRICE/UNIT	AMOUNT

SPECIAL INSTRUCTION

_____
_____
_____

ESTIMATE NOTES | DIAGRAMS

ESTIMATE FORM

DATE: JOB CATEGORY: REFERRED BY:

CLIENT NAME: PHONE:

ADDRESS: EMAIL:

ENQUIRY MODE ◯ PHONE ◯ EMAIL ◯ ONLINE ◯ PHYSICAL | OTHER

HOW DID YOU KNOW ABOUT US:

APPOINTMENT DATE & DATE:

WORK DESCRIPTION

_____

_____

_____

_____

_____

_____

JOB ACCEPTED? [ ] YES [ ] NO   REASON FOR DECLINE:

ESTIMATED JOB TIME/DAYS: ACTUAL FINISH DATE: REASON FOR DELAY:

TOTAL HRS/DAYS?:   RATE/HR/DAY: SCHEDULED INSTALL DATE:

QUOTE PRICE: AGREED PRICE: PAYMENT DATE: PAYMENT MODE:

DELAY? [ ] YES [ ] NO   REASON: RESCHEDULED DATE:

MATERIALS NEEDED

MATERIAL	QTY	PRICE/UNIT	AMOUNT	MATERIAL	QTY	PRICE/UNIT	AMOUNT

SPECIAL INSTRUCTION

_____

_____

_____

ESTIMATE NOTES | DIAGRAMS

ESTIMATE FORM

DATE: _____ JOB CATEGORY: _____ REFERRED BY: _____

CLIENT NAME: _____ PHONE: _____

ADDRESS: _____ EMAIL: _____

ENQUIRY MODE ◯ PHONE ◯ EMAIL ◯ ONLINE ◯ PHYSICAL | OTHER _____

HOW DID YOU KNOW ABOUT US: _____

APPOINTMENT DATE & DATE: _____ | _____ | _____

WORK DESCRIPTION

_____
_____
_____
_____
_____
_____
_____

JOB ACCEPTED? [ ] YES [ ] NO REASON FOR DECLINE: _____

ESTIMATED JOB TIME/DAYS: _____ ACTUAL FINISH DATE: _____ REASON FOR DELAY: _____

TOTAL HRS/DAYS?: _____ RATE/HR/DAY: _____ SCHEDULED INSTALL DATE: _____

QUOTE PRICE: _____ AGREED PRICE: _____ PAYMENT DATE: _____ PAYMENT MODE: _____

DELAY? [ ] YES [ ] NO REASON: _____ RESCHEDULED DATE: _____

MATERIALS NEEDED

MATERIAL	QTY	PRICE/UNIT	AMOUNT	MATERIAL	QTY	PRICE/UNIT	AMOUNT

SPECIAL INSTRUCTION

_____
_____
_____

ESTIMATE NOTES | DIAGRAMS

ESTIMATE FORM

DATE: _____ JOB CATEGORY: _____ REFERRED BY: _____

CLIENT NAME: _____ PHONE: _____

ADDRESS: _____ EMAIL: _____

ENQUIRY MODE ◯ PHONE ◯ EMAIL ◯ ONLINE ◯ PHYSICAL | OTHER _____

HOW DID YOU KNOW ABOUT US: _____

APPOINTMENT DATE & DATE: _____

WORK DESCRIPTION

_____
_____
_____
_____
_____
_____

JOB ACCEPTED? [ ] YES [ ] NO REASON FOR DECLINE: _____

ESTIMATED JOB TIME/DAYS: _____ ACTUAL FINISH DATE: _____ REASON FOR DELAY: _____

TOTAL HRS/DAYS?: _____ RATE/HR/DAY: _____ SCHEDULED INSTALL DATE: _____

QUOTE PRICE: _____ AGREED PRICE: _____ PAYMENT DATE: _____ PAYMENT MODE: _____

DELAY? [ ] YES [ ] NO REASON: _____ RESCHEDULED DATE: _____

MATERIALS NEEDED

MATERIAL	QTY	PRICE/UNIT	AMOUNT	MATERIAL	QTY	PRICE/UNIT	AMOUNT

SPECIAL INSTRUCTION

_____
_____
_____

ESTIMATE NOTES | DIAGRAMS

ESTIMATE FORM

DATE: | JOB CATEGORY: | REFERRED BY:

CLIENT NAME: | PHONE:

ADDRESS: | EMAIL:

ENQUIRY MODE ◯ PHONE ◯ EMAIL ◯ ONLINE ◯ PHYSICAL | OTHER

HOW DID YOU KNOW ABOUT US:

APPOINTMENT DATE & DATE: | |

WORK DESCRIPTION

_____
_____
_____
_____
_____
_____

JOB ACCEPTED? [ ] YES [ ] NO REASON FOR DECLINE:

ESTIMATED JOB TIME/DAYS: | ACTUAL FINISH DATE: | REASON FOR DELAY:

TOTAL HRS/DAYS?: | RATE/HR/DAY: | SCHEDULED INSTALL DATE:

QUOTE PRICE: | AGREED PRICE: | PAYMENT DATE: | PAYMENT MODE:

DELAY? [ ] YES [ ] NO REASON: | RESCHEDULED DATE:

MATERIALS NEEDED

MATERIAL	QTY	PRICE/UNIT	AMOUNT	MATERIAL	QTY	PRICE/UNIT	AMOUNT

SPECIAL INSTRUCTION

_____
_____
_____

ESTIMATE NOTES | DIAGRAMS

ESTIMATE FORM

DATE: | JOB CATEGORY: | REFERRED BY:

CLIENT NAME: | PHONE:

ADDRESS: | EMAIL:

ENQUIRY MODE ◯ PHONE ◯ EMAIL ◯ ONLINE ◯ PHYSICAL | OTHER

HOW DID YOU KNOW ABOUT US:

APPOINTMENT DATE & DATE:

WORK DESCRIPTION

_____
_____
_____
_____
_____
_____

JOB ACCEPTED? [ ] YES [ ] NO   REASON FOR DECLINE:

ESTIMATED JOB TIME/DAYS: | ACTUAL FINISH DATE: | REASON FOR DELAY:

TOTAL HRS/DAYS?: | RATE/HR/DAY: | SCHEDULED INSTALL DATE:

QUOTE PRICE: | AGREED PRICE: | PAYMENT DATE: | PAYMENT MODE:

DELAY? [ ] YES   [ ] NO   REASON: | RESCHEDULED DATE:

MATERIALS NEEDED

MATERIAL	QTY	PRICE/UNIT	AMOUNT	MATERIAL	QTY	PRICE/UNIT	AMOUNT

SPECIAL INSTRUCTION

_____
_____
_____

ESTIMATE NOTES | DIAGRAMS

ESTIMATE FORM

DATE: JOB CATEGORY: REFERRED BY:

CLIENT NAME: PHONE:

ADDRESS: EMAIL:

ENQUIRY MODE   ◯ PHONE   ◯ EMAIL   ◯ ONLINE   ◯ PHYSICAL | OTHER

HOW DID YOU KNOW ABOUT US:

APPOINTMENT DATE & DATE: | |

WORK DESCRIPTION

_____
_____
_____
_____
_____
_____
_____

JOB ACCEPTED? [ ] YES [ ] NO   REASON FOR DECLINE:

ESTIMATED JOB TIME/DAYS: ACTUAL FINISH DATE: REASON FOR DELAY:

TOTAL HRS/DAYS?:   RATE/HR/DAY: SCHEDULED INSTALL DATE:

QUOTE PRICE: AGREED PRICE: PAYMENT DATE: PAYMENT MODE:

DELAY? [ ] YES [ ] NO   REASON: RESCHEDULED DATE:

MATERIALS NEEDED

MATERIAL	QTY	PRICE/UNIT	AMOUNT	MATERIAL	QTY	PRICE/UNIT	AMOUNT

SPECIAL INSTRUCTION

_____
_____
_____

ESTIMATE NOTES | DIAGRAMS

ESTIMATE FORM

DATE: JOB CATEGORY: REFERRED BY:

CLIENT NAME: PHONE:

ADDRESS: EMAIL:

ENQUIRY MODE ◯ PHONE ◯ EMAIL ◯ ONLINE ◯ PHYSICAL | OTHER

HOW DID YOU KNOW ABOUT US:

APPOINTMENT DATE & DATE:

WORK DESCRIPTION

_____
_____
_____
_____
_____
_____

JOB ACCEPTED? [ ] YES [ ] NO   REASON FOR DECLINE:

ESTIMATED JOB TIME/DAYS: ACTUAL FINISH DATE: REASON FOR DELAY:

TOTAL HRS/DAYS?:   RATE/HR/DAY: SCHEDULED INSTALL DATE:

QUOTE PRICE: AGREED PRICE: PAYMENT DATE: PAYMENT MODE:

DELAY? [ ] YES [ ] NO   REASON: RESCHEDULED DATE:

MATERIALS NEEDED

MATERIAL	QTY	PRICE/UNIT	AMOUNT	MATERIAL	QTY	PRICE/UNIT	AMOUNT

SPECIAL INSTRUCTION

_____
_____
_____

ESTIMATE NOTES | DIAGRAMS

ESTIMATE FORM

DATE: 　　　　　JOB CATEGORY: 　　　　　REFERRED BY:

CLIENT NAME: 　　　　　PHONE:

ADDRESS: 　　　　　EMAIL:

ENQUIRY MODE ◯ PHONE ◯ EMAIL ◯ ONLINE ◯ PHYSICAL | OTHER

HOW DID YOU KNOW ABOUT US:

APPOINTMENT DATE & DATE:

WORK DESCRIPTION

_____
_____
_____
_____
_____
_____

JOB ACCEPTED? [ ] YES [ ] NO REASON FOR DECLINE:

ESTIMATED JOB TIME/DAYS: 　　　　ACTUAL FINISH DATE: 　　　　REASON FOR DELAY:

TOTAL HRS/DAYS?: 　　　　RATE/HR/DAY: 　　　　SCHEDULED INSTALL DATE:

QUOTE PRICE: 　　　　AGREED PRICE: 　　　　PAYMENT DATE: 　　　　PAYMENT MODE:

DELAY? [ ] YES [ ] NO REASON: 　　　　RESCHEDULED DATE:

MATERIALS NEEDED

MATERIAL	QTY	PRICE/UNIT	AMOUNT	MATERIAL	QTY	PRICE/UNIT	AMOUNT

SPECIAL INSTRUCTION

_____
_____
_____

ESTIMATE NOTES | DIAGRAMS

ESTIMATE FORM

DATE: JOB CATEGORY: REFERRED BY:

CLIENT NAME: PHONE:

ADDRESS: EMAIL:

ENQUIRY MODE ◯ PHONE ◯ EMAIL ◯ ONLINE ◯ PHYSICAL | OTHER

HOW DID YOU KNOW ABOUT US:

APPOINTMENT DATE & DATE:

WORK DESCRIPTION

_____
_____
_____
_____
_____
_____
_____

JOB ACCEPTED? [ ] YES [ ] NO   REASON FOR DECLINE:

ESTIMATED JOB TIME/DAYS:	ACTUAL FINISH DATE:	REASON FOR DELAY:
TOTAL HRS/DAYS?:	RATE/HR/DAY:	SCHEDULED INSTALL DATE:
QUOTE PRICE:	AGREED PRICE:	PAYMENT DATE:	PAYMENT MODE:
DELAY? [ ] YES [ ] NO REASON:		RESCHEDULED DATE:

MATERIALS NEEDED

MATERIAL	QTY	PRICE/UNIT	AMOUNT	MATERIAL	QTY	PRICE/UNIT	AMOUNT

SPECIAL INSTRUCTION

_____
_____
_____

ESTIMATE NOTES | DIAGRAMS

ESTIMATE FORM

DATE: | JOB CATEGORY: | REFERRED BY:

CLIENT NAME: | PHONE:

ADDRESS: | EMAIL:

ENQUIRY MODE ◯ PHONE ◯ EMAIL ◯ ONLINE ◯ PHYSICAL | OTHER

HOW DID YOU KNOW ABOUT US:

APPOINTMENT DATE & DATE: | |

WORK DESCRIPTION

_____
_____
_____
_____
_____
_____

JOB ACCEPTED? [ ] YES [ ] NO   REASON FOR DECLINE:

ESTIMATED JOB TIME/DAYS: | ACTUAL FINISH DATE: | REASON FOR DELAY:

TOTAL HRS/DAYS?: | RATE/HR/DAY: | SCHEDULED INSTALL DATE:

QUOTE PRICE: | AGREED PRICE: | PAYMENT DATE: | PAYMENT MODE:

DELAY? [ ] YES [ ] NO   REASON: | RESCHEDULED DATE:

MATERIALS NEEDED

MATERIAL	QTY	PRICE/UNIT	AMOUNT	MATERIAL	QTY	PRICE/UNIT	AMOUNT

SPECIAL INSTRUCTION

_____
_____
_____

ESTIMATE NOTES | DIAGRAMS

ESTIMATE FORM

DATE: | JOB CATEGORY: | REFERRED BY:

CLIENT NAME: | PHONE:

ADDRESS: | EMAIL:

ENQUIRY MODE ◯ PHONE ◯ EMAIL ◯ ONLINE ◯ PHYSICAL | OTHER

HOW DID YOU KNOW ABOUT US:

APPOINTMENT DATE & DATE:

WORK DESCRIPTION

_____
_____
_____
_____
_____
_____

JOB ACCEPTED? [ ] YES [ ] NO REASON FOR DECLINE:

ESTIMATED JOB TIME/DAYS: | ACTUAL FINISH DATE: | REASON FOR DELAY:

TOTAL HRS/DAYS?: | RATE/HR/DAY: | SCHEDULED INSTALL DATE:

QUOTE PRICE: | AGREED PRICE: | PAYMENT DATE: | PAYMENT MODE:

DELAY? [ ] YES [ ] NO REASON: | RESCHEDULED DATE:

MATERIALS NEEDED

MATERIAL	QTY	PRICE/UNIT	AMOUNT	MATERIAL	QTY	PRICE/UNIT	AMOUNT

SPECIAL INSTRUCTION

_____
_____
_____

ESTIMATE NOTES | DIAGRAMS

WORK SCHEDULE

MONTH:

YEAR:

SUNDAY				
MONDAY				
TUESDAY				
WEDNESDAY				
THURSDAY				
FRIDAY				
SATURDAY				

WORK SCHEDULE

MONTH:

YEAR:

SUNDAY				
MONDAY				
TUESDAY				
WEDNESDAY				
THURSDAY				
FRIDAY				
SATURDAY				

WORK SCHEDULE

MONTH:

YEAR:

SUNDAY				
MONDAY				
TUESDAY				
WEDNESDAY				
THURSDAY				
FRIDAY				
SATURDAY				

WORK SCHEDULE

MONTH:

YEAR:

SUNDAY				
MONDAY				
TUESDAY				
WEDNESDAY				
THURSDAY				
FRIDAY				
SATURDAY				

WORK SCHEDULE

MONTH:

YEAR:

	SUNDAY	MONDAY	TUESDAY	WEDNESDAY	THURSDAY	FRIDAY	SATURDAY

WORK SCHEDULE

MONTH: **YEAR:**

SUNDAY				
MONDAY				
TUESDAY				
WEDNESDAY				
THURSDAY				
FRIDAY				
SATURDAY				

WORK SCHEDULE

MONTH: **YEAR:**

	SUNDAY	MONDAY	TUESDAY	WEDNESDAY	THURSDAY	FRIDAY	SATURDAY

WORK SCHEDULE

MONTH: YEAR:

SUNDAY				
MONDAY				
TUESDAY				
WEDNESDAY				
THURSDAY				
FRIDAY				
SATURDAY				

WORK SCHEDULE

MONTH:

YEAR:

SUNDAY				
MONDAY				
TUESDAY				
WEDNESDAY				
THURSDAY				
FRIDAY				
SATURDAY				

WORK SCHEDULE

MONTH:

YEAR:

SUNDAY				
MONDAY				
TUESDAY				
WEDNESDAY				
THURSDAY				
FRIDAY				
SATURDAY				

WORK SCHEDULE

MONTH: **YEAR:**

SUNDAY				
MONDAY				
TUESDAY				
WEDNESDAY				
THURSDAY				
FRIDAY				
SATURDAY				

WORK SCHEDULE

MONTH:

YEAR:

SUNDAY				
MONDAY				
TUESDAY				
WEDNESDAY				
THURSDAY				
FRIDAY				
SATURDAY				

WORK SCHEDULE

MONTH:

YEAR:

	SUNDAY			
MONDAY				
TUESDAY				
WEDNESDAY				
THURSDAY				
FRIDAY				
SATURDAY				